Elementary Statistics Tables

Henry R. Neave
University of Nottingham

London
GEORGE ALLEN & UNWIN
Boston   Sydney

Preface

Having published my *Statistics tables* in 1978, the obvious question is: why another book of Statistics tables so soon afterwards? The answer derives from reactions to the first book from a sample of some 500 lecturers and teachers covering a wide range both of educational establishments and of departments within those establishments. Approximately half found *Statistics tables* suitable for their needs; however the other half indicated that their courses covered rather less topics than included in the Tables, and therefore that a less comprehensive collection would be adequate. Further, some North American advisers suggested that more 'on the spot' descriptions, directions and illustrative examples would make such a book far more attractive and useful. *Elementary statistics tables* has been produced with these comments very much in mind.

The coverage of topics is probably still wider than in most introductory Statistics courses. But useful techniques are often omitted from such courses because of the lack of good tables or charts in the textbook being used, and it is one of the aims of this book to enable instructors to broaden the range of statistical methods included in their syllabuses. Even if some of the methods are completely omitted from the course textbook, instructors and students will find that these pages contain brief but adequate explanations and illustrations.

In deciding the topics to be included, I was guided to an extent by draft proposals for the Technician Education Council (TEC) awards, and *Elementary statistics tables* essentially covers the areas included in this scheme for which tables and/or charts are necessary. The standard distributions are of course included, i.e. binomial, Poisson, normal, t, χ^2 and F. Both individual and cumulative probabilities are given for binomial and Poisson distributions, the cumulative Poisson probabilities being derived from a newly designed chart on which the curves are virtually straight: this should enhance ease of reading and accuracy. A selection of useful nonparametric techniques is included, and advocates of these excellent and easy-to-apply methods will notice the inclusion of considerably improved tables for the Kruskal–Wallis and Friedman tests, and a new table for a Kolmogorov–Smirnov general test for normality. The book also contains random-number tables, including random numbers from normal and exponential distributions (useful for simple simulation experiments), binomial coefficients, control chart constants, various tables and charts concerned with correlation and rank correlation, and charts giving confidence intervals for a binomial p. The book ends with four pages of familiar mathematical tables and a table of useful constants, and a glossary of symbols used in the book will be found inside the back cover.

Considerable care and thought has been given to the design and layout of the tables. Special care has been taken to simplify a matter which many students find confusing: which table entries to use for one-sided and two-sided tests and for confidence intervals. Several tables, such as the percentage points for the normal, t, χ^2 and F distributions, may be used for several purposes. Throughout this book, α_1 and α_2 are used to denote significance levels for one-sided (or 'one-tailed') and two-sided tests, respectively, and γ indicates confidence levels for confidence intervals. (Where occasion demands, we even go so far as to use α_1^R and α_1^L to denote significance levels for right-hand and left-hand one-sided tests.) If a table can be used for all three purposes, all three cases are clearly indicated, with 5% and 1% critical values and 95% and 99% confidence levels being highlighted.

My thanks are due to many people who have contributed in various ways to the production of this book. I am especially grateful to Peter Worthington and Arthur Morley for their help and guidance throughout its development: Peter deserves special mention for his large contribution to the new tables for the Kruskal–Wallis and Friedman tests. Thanks also to Graham Littler and John Silk who very usefully reviewed some early proposals, and to Trevor Easingwood for discussions concerning the TEC proposals. At the time of writing, the proof-reading stage has not yet arrived; but thanks in advance to Tonie-Carol Brown who will be helping me with that unenviable task. Finally, I must express my gratitude to the staff of the Cripps Computing Centre at Nottingham University: all of the tables and charts have been newly computed for this publication, and the service which they have provided has been excellent.

Naturally, total responsibility for any errors is mine alone. It would be nice to think that there are none, but I would greatly appreciate anybody who sees anything that they know or suspect to be incorrect communicating the facts immediately to me.

HENRY NEAVE
October 1979

Contents

George Allen & Unwin (Publishers) Ltd,
40 Museum Street, London WC1A 1LU, UK

George Allen & Unwin (Publishers) Ltd,
Park Lane, Hemel Hempstead, Herts HP2 4TE, UK

Allen & Unwin Inc.,
9 Winchester Terrace, Winchester, Mass. 01890, USA

George Allen & Unwin Australia Pty Ltd,
8 Napier Street, North Sydney, NSW 2060, Australia

First published in 1981
Second impression, 1983

British Library Cataloguing in Publication Data

Neave, Henry Robert
  Elementary statistics tables.
1. Mathematical statistics - Tables, etc.
I. Title
519.5'021'2   QA276.25   80-40430
ISBN 0-04-001002-3

Set in Press Roman by Alden Press, Oxford, London and Northampton and printed in Great Britain by Biddles Ltd, Guildford, Surrey.

The binomial distribution: individual probabilities

$$\text{Prob}(X = x) = \binom{n}{x} p^x (1-p)^{n-x} \qquad (x = 0, 1, \ldots, n)$$

Prob $(X = x)$

p	.01	.02	.03	.04	.05	.06	.07	.08	.09	.10	.15	$\frac{1}{6}$	.20	.25	.30	$\frac{1}{3}$	.35	.40	.45	.50	

n = 1

x	.01	.02	.03	.04	.05	.06	.07	.08	.09	.10	.15	$\frac{1}{6}$	.20	.25	.30	$\frac{1}{3}$	.35	.40	.45	.50	
0	.9900	.9800	.9700	.9600	.9500	.9400	.9300	.9200	.9100	.9000	.8500	.8333	.8000	.7500	.7000	.6667	.6500	.6000	.5500	.5000	1
1	.0100	.0200	.0300	.0400	.0500	.0600	.0700	.0800	.0900	.1000	.1500	.1667	.2000	.2500	.3000	.3333	.3500	.4000	.4500	.5000	0

n = 2

x	.01	.02	.03	.04	.05	.06	.07	.08	.09	.10	.15	$\frac{1}{6}$	.20	.25	.30	$\frac{1}{3}$	.35	.40	.45	.50	
0	.9801	.9604	.9409	.9216	.9025	.8836	.8649	.8464	.8281	.8100	.7225	.6944	.6400	.5625	.4900	.4444	.4225	.3600	.3025	.2500	2
1	.0198	.0392	.0582	.0768	.0950	.1128	.1302	.1472	.1638	.1800	.2550	.2778	.3200	.3750	.4200	.4444	.4550	.4800	.4950	.5000	1
2	.0001	.0004	.0009	.0016	.0025	.0036	.0049	.0064	.0081	.0100	.0225	.0278	.0400	.0625	.0900	.1111	.1225	.1600	.2025	.2500	0

n = 3

x	.01	.02	.03	.04	.05	.06	.07	.08	.09	.10	.15	$\frac{1}{6}$	.20	.25	.30	$\frac{1}{3}$	.35	.40	.45	.50	
0	.9703	.9412	.9127	.8847	.8574	.8306	.8044	.7787	.7536	.7290	.6141	.5787	.5120	.4219	.3430	.2963	.2746	.2160	.1664	.1250	3
1	.0294	.0576	.0847	.1106	.1354	.1590	.1816	.2031	.2236	.2430	.3251	.3472	.3840	.4219	.4410	.4444	.4436	.4320	.4084	.3750	2
2	.0003	.0012	.0026	.0046	.0071	.0102	.0137	.0177	.0221	.0270	.0574	.0694	.0960	.1406	.1890	.2222	.2389	.2880	.3341	.3750	1
3	.0000	.0000	.0000	.0001	.0001	.0002	.0003	.0005	.0007	.0010	.0034	.0046	.0080	.0156	.0270	.0370	.0429	.0640	.0911	.1250	0

n = 4

x	.01	.02	.03	.04	.05	.06	.07	.08	.09	.10	.15	$\frac{1}{6}$	.20	.25	.30	$\frac{1}{3}$	.35	.40	.45	.50	
0	.9606	.9224	.8853	.8493	.8145	.7807	.7481	.7164	.6857	.6561	.5220	.4823	.4096	.3164	.2401	.1975	.1785	.1296	.0915	.0625	4
1	.0388	.0753	.1095	.1416	.1715	.1993	.2252	.2492	.2713	.2916	.3685	.3858	.4096	.4219	.4116	.3951	.3845	.3456	.2995	.2500	3
2	.0006	.0023	.0051	.0088	.0135	.0191	.0254	.0325	.0402	.0486	.0975	.1157	.1536	.2109	.2646	.2963	.3105	.3456	.3675	.3750	2
3	.0000	.0000	.0001	.0002	.0005	.0008	.0013	.0019	.0027	.0036	.0115	.0154	.0256	.0469	.0756	.0988	.1115	.1536	.2005	.2500	1
4	.0000	.0000	.0000	.0000	.0000	.0000	.0000	.0000	.0001	.0001	.0005	.0008	.0016	.0039	.0081	.0123	.0150	.0256	.0410	.0625	0

n = 5

x	.01	.02	.03	.04	.05	.06	.07	.08	.09	.10	.15	$\frac{1}{6}$	.20	.25	.30	$\frac{1}{3}$	.35	.40	.45	.50	
0	.9510	.9039	.8587	.8154	.7738	.7339	.6957	.6591	.6240	.5905	.4437	.4019	.3277	.2373	.1681	.1317	.1160	.0778	.0503	.0313	5
1	.0480	.0922	.1328	.1699	.2036	.2342	.2618	.2866	.3086	.3281	.3915	.4019	.4096	.3955	.3602	.3292	.3124	.2592	.2059	.1563	4
2	.0010	.0038	.0082	.0142	.0214	.0299	.0394	.0498	.0610	.0729	.1382	.1608	.2048	.2637	.3087	.3292	.3364	.3456	.3369	.3125	3
3	.0000	.0001	.0003	.0006	.0011	.0019	.0030	.0043	.0060	.0081	.0244	.0322	.0512	.0879	.1323	.1646	.1811	.2304	.2757	.3125	2
4	.0000	.0000	.0000	.0000	.0000	.0001	.0001	.0002	.0003	.0005	.0022	.0032	.0064	.0146	.0284	.0412	.0488	.0768	.1128	.1563	1
5	.0000	.0000	.0000	.0000	.0000	.0000	.0000	.0000	.0000	.0000	.0001	.0001	.0003	.0010	.0024	.0041	.0053	.0102	.0185	.0313	0

n = 6

x	.01	.02	.03	.04	.05	.06	.07	.08	.09	.10	.15	$\frac{1}{6}$	.20	.25	.30	$\frac{1}{3}$	.35	.40	.45	.50	
0	.9415	.8858	.8330	.7828	.7351	.6899	.6470	.6064	.5679	.5314	.3771	.3349	.2621	.1780	.1176	.0878	.0754	.0467	.0277	.0156	6
1	.0571	.1085	.1546	.1957	.2321	.2642	.2922	.3164	.3370	.3543	.3993	.4019	.3932	.3560	.3025	.2634	.2437	.1866	.1359	.0938	5
2	.0014	.0055	.0120	.0204	.0305	.0422	.0550	.0688	.0833	.0984	.1762	.2009	.2458	.2966	.3241	.3292	.3280	.3110	.2780	.2344	4
3	.0000	.0002	.0005	.0011	.0021	.0036	.0055	.0080	.0110	.0146	.0415	.0536	.0819	.1318	.1852	.2195	.2355	.2765	.3032	.3125	3
4	.0000	.0000	.0000	.0000	.0001	.0002	.0003	.0005	.0008	.0012	.0055	.0080	.0154	.0330	.0595	.0823	.0951	.1382	.1861	.2344	2
5	.0000	.0000	.0000	.0000	.0000	.0000	.0000	.0000	.0000	.0001	.0004	.0006	.0015	.0044	.0102	.0165	.0205	.0369	.0609	.0938	1
6	.0000	.0000	.0000	.0000	.0000	.0000	.0000	.0000	.0000	.0000	.0000	.0000	.0001	.0002	.0007	.0014	.0018	.0041	.0083	.0156	0

n = 7

x	.01	.02	.03	.04	.05	.06	.07	.08	.09	.10	.15	$\frac{1}{6}$	.20	.25	.30	$\frac{1}{3}$	.35	.40	.45	.50	
0	.9321	.8681	.8080	.7514	.6983	.6485	.6017	.5578	.5168	.4783	.3206	.2791	.2097	.1335	.0824	.0585	.0490	.0280	.0152	.0078	7
1	.0659	.1240	.1749	.2192	.2573	.2897	.3170	.3396	.3578	.3720	.3960	.3907	.3670	.3115	.2471	.2048	.1848	.1306	.0872	.0547	6
2	.0020	.0076	.0162	.0274	.0406	.0555	.0716	.0886	.1061	.1240	.2097	.2344	.2753	.3115	.3177	.3073	.2985	.2613	.2140	.1641	5
3	.0000	.0003	.0008	.0019	.0036	.0059	.0090	.0128	.0175	.0230	.0617	.0781	.1147	.1730	.2269	.2561	.2679	.2903	.2918	.2734	4
4	.0000	.0000	.0000	.0001	.0002	.0004	.0007	.0011	.0017	.0026	.0109	.0156	.0287	.0577	.0972	.1280	.1442	.1935	.2388	.2734	3
5	.0000	.0000	.0000	.0000	.0000	.0000	.0000	.0001	.0001	.0002	.0012	.0019	.0043	.0115	.0250	.0384	.0466	.0774	.1172	.1641	2
6	.0000	.0000	.0000	.0000	.0000	.0000	.0000	.0000	.0000	.0000	.0001	.0001	.0004	.0013	.0036	.0064	.0084	.0172	.0320	.0547	1
7	.0000	.0000	.0000	.0000	.0000	.0000	.0000	.0000	.0000	.0000	.0000	.0000	.0000	.0001	.0002	.0005	.0006	.0016	.0037	.0078	0

n = 8

x	.01	.02	.03	.04	.05	.06	.07	.08	.09	.10	.15	$\frac{1}{6}$	.20	.25	.30	$\frac{1}{3}$	.35	.40	.45	.50	
0	.9227	.8508	.7837	.7214	.6634	.6096	.5596	.5132	.4703	.4305	.2725	.2326	.1678	.1001	.0576	.0390	.0319	.0168	.0084	.0039	8
1	.0746	.1389	.1939	.2405	.2793	.3113	.3370	.3570	.3721	.3826	.3847	.3721	.3355	.2670	.1977	.1561	.1373	.0896	.0548	.0313	7
2	.0026	.0099	.0210	.0351	.0515	.0695	.0888	.1087	.1288	.1488	.2376	.2605	.2936	.3115	.2965	.2731	.2587	.2090	.1569	.1094	6
3	.0001	.0004	.0013	.0029	.0054	.0089	.0134	.0189	.0255	.0331	.0839	.1042	.1468	.2076	.2541	.2731	.2786	.2787	.2568	.2188	5
4	.0000	.0000	.0001	.0002	.0004	.0007	.0013	.0021	.0031	.0046	.0185	.0260	.0459	.0865	.1361	.1707	.1875	.2322	.2627	.2734	4
5	.0000	.0000	.0000	.0000	.0000	.0000	.0001	.0001	.0002	.0004	.0026	.0042	.0092	.0231	.0467	.0683	.0808	.1239	.1719	.2188	3
6	.0000	.0000	.0000	.0000	.0000	.0000	.0000	.0000	.0000	.0000	.0002	.0004	.0011	.0038	.0100	.0171	.0217	.0413	.0703	.1094	2
7	.0000	.0000	.0000	.0000	.0000	.0000	.0000	.0000	.0000	.0000	.0000	.0000	.0001	.0004	.0012	.0024	.0033	.0079	.0164	.0313	1
8	.0000	.0000	.0000	.0000	.0000	.0000	.0000	.0000	.0000	.0000	.0000	.0000	.0000	.0000	.0001	.0002	.0002	.0007	.0017	.0039	0

p	.99	.98	.97	.96	.95	.94	.93	.92	.91	.90	.85	$\frac{5}{6}$	.80	.75	.70	$\frac{2}{3}$	.65	.60	.55	.50	p

Prob $(X = x)$

If the probability is p that a certain event (often called a 'success') occurs in a trial of an experiment, the binomial distribution is concerned with the total number X of successes obtained in n independent trials of the experiment. Pages 4, 6, 8 and 10 give Prob $(X = x)$ for all possible x and n up to 20, and 39 values of p. For values of $p \leqslant \frac{1}{2}$ (along the top horizontal) refer to the x-values in the left-hand column; for values of $p \geqslant \frac{1}{2}$ (along the bottom horizontal) refer to the x-values in the right-hand column.

The binomial distribution: cumulative probabilities

$$\text{Prob}\,(X \geqslant x) = \sum_{r=x}^{n} \binom{n}{r} p^r (1-p)^{n-r} \quad \text{for } p \leqslant \tfrac{1}{2} \qquad \text{Prob}\,(X \leqslant x) = \sum_{r=0}^{x} \binom{n}{r} p^r (1-p)^{n-r} \quad \text{for } p \geqslant \tfrac{1}{2}$$

Prob $(X \geqslant x)$

p	.01	.02	.03	.04	.05	.06	.07	.08	.09	.10	.15	1/6	.20	.25	.30	1/3	.35	.40	.45	.50	

n = 1

x	.01	.02	.03	.04	.05	.06	.07	.08	.09	.10	.15	1/6	.20	.25	.30	1/3	.35	.40	.45	.50	
0	1.000	1.000	1.000	1.000	1.000	1.000	1.000	1.000	1.000	1.000	1.000	1.000	1.000	1.000	1.000	1.000	1.000	1.000	1.000	1.000	1
1	.0100	.0200	.0300	.0400	.0500	.0600	.0700	.0800	.0900	.1000	.1500	.1667	.2000	.2500	.3000	.3333	.3500	.4000	.4500	.5000	0

n = 2

x	.01	.02	.03	.04	.05	.06	.07	.08	.09	.10	.15	1/6	.20	.25	.30	1/3	.35	.40	.45	.50	
0	1.000	1.000	1.000	1.000	1.000	1.000	1.000	1.000	1.000	1.000	1.000	1.000	1.000	1.000	1.000	1.000	1.000	1.000	1.000	1.000	2
1	.0199	.0396	.0591	.0784	.0975	.1164	.1351	.1536	.1719	.1900	.2775	.3056	.3600	.4375	.5100	.5556	.5775	.6400	.6975	.7500	1
2	.0001	.0004	.0009	.0016	.0025	.0036	.0049	.0064	.0081	.0100	.0225	.0278	.0400	.0625	.0900	.1111	.1225	.1600	.2025	.2500	0

n = 3

x	.01	.02	.03	.04	.05	.06	.07	.08	.09	.10	.15	1/6	.20	.25	.30	1/3	.35	.40	.45	.50	
0	1.000	1.000	1.000	1.000	1.000	1.000	1.000	1.000	1.000	1.000	1.000	1.000	1.000	1.000	1.000	1.000	1.000	1.000	1.000	1.000	3
1	.0297	.0588	.0873	.1153	.1426	.1694	.1956	.2213	.2464	.2710	.3859	.4213	.4880	.5781	.6570	.7037	.7254	.7840	.8336	.8750	2
2	.0003	.0012	.0026	.0047	.0073	.0104	.0140	.0182	.0228	.0280	.0608	.0741	.1040	.1563	.2160	.2593	.2818	.3520	.4252	.5000	1
3	.0000	.0000	.0000	.0001	.0001	.0002	.0003	.0005	.0007	.0010	.0034	.0046	.0080	.0156	.0270	.0370	.0429	.0640	.0911	.1250	0

n = 4

x	.01	.02	.03	.04	.05	.06	.07	.08	.09	.10	.15	1/6	.20	.25	.30	1/3	.35	.40	.45	.50	
0	1.000	1.000	1.000	1.000	1.000	1.000	1.000	1.000	1.000	1.000	1.000	1.000	1.000	1.000	1.000	1.000	1.000	1.000	1.000	1.000	4
1	.0394	.0776	.1147	.1507	.1855	.2193	.2519	.2836	.3143	.3439	.4780	.5177	.5904	.6836	.7599	.8025	.8215	.8704	.9085	.9375	3
2	.0006	.0023	.0052	.0091	.0140	.0199	.0267	.0344	.0430	.0523	.1095	.1319	.1808	.2617	.3483	.4074	.4370	.5248	.6090	.6875	2
3	.0000	.0000	.0001	.0002	.0005	.0008	.0013	.0019	.0027	.0037	.0120	.0162	.0272	.0508	.0837	.1111	.1265	.1792	.2415	.3125	1
4	.0000	.0000	.0000	.0000	.0000	.0000	.0000	.0000	.0001	.0001	.0005	.0008	.0016	.0039	.0081	.0123	.0150	.0256	.0410	.0625	0

n = 5

x	.01	.02	.03	.04	.05	.06	.07	.08	.09	.10	.15	1/6	.20	.25	.30	1/3	.35	.40	.45	.50	
0	1.000	1.000	1.000	1.000	1.000	1.000	1.000	1.000	1.000	1.000	1.000	1.000	1.000	1.000	1.000	1.000	1.000	1.000	1.000	1.000	5
1	.0490	.0961	.1413	.1846	.2262	.2661	.3043	.3409	.3760	.4095	.5563	.5981	.6723	.7627	.8319	.8683	.8840	.9222	.9497	.9688	4
2	.0010	.0038	.0085	.0148	.0226	.0319	.0425	.0544	.0674	.0815	.1648	.1962	.2627	.3672	.4718	.5391	.5716	.6630	.7438	.8125	3
3	.0000	.0001	.0003	.0006	.0012	.0020	.0031	.0045	.0063	.0086	.0266	.0355	.0579	.1035	.1631	.2099	.2352	.3174	.4069	.5000	2
4	.0000	.0000	.0000	.0000	.0001	.0001	.0001	.0002	.0003	.0005	.0022	.0033	.0067	.0156	.0308	.0453	.0540	.0870	.1312	.1875	1
5	.0000	.0000	.0000	.0000	.0000	.0000	.0000	.0000	.0000	.0000	.0001	.0001	.0003	.0010	.0024	.0041	.0053	.0102	.0185	.0313	0

n = 6

x	.01	.02	.03	.04	.05	.06	.07	.08	.09	.10	.15	1/6	.20	.25	.30	1/3	.35	.40	.45	.50	
0	1.000	1.000	1.000	1.000	1.000	1.000	1.000	1.000	1.000	1.000	1.000	1.000	1.000	1.000	1.000	1.000	1.000	1.000	1.000	1.000	6
1	.0585	.1142	.1670	.2172	.2649	.3101	.3530	.3936	.4321	.4686	.6229	.6651	.7379	.8220	.8824	.9122	.9246	.9533	.9723	.9844	5
2	.0015	.0057	.0125	.0216	.0328	.0459	.0608	.0773	.0952	.1143	.2235	.2632	.3446	.4661	.5798	.6488	.6809	.7667	.8364	.8906	4
3	.0000	.0002	.0005	.0012	.0022	.0038	.0058	.0085	.0118	.0158	.0473	.0623	.0989	.1694	.2557	.3196	.3529	.4557	.5585	.6563	3
4	.0000	.0000	.0000	.0000	.0001	.0002	.0003	.0005	.0008	.0013	.0059	.0087	.0170	.0376	.0705	.1001	.1174	.1792	.2553	.3438	2
5	.0000	.0000	.0000	.0000	.0000	.0000	.0000	.0000	.0000	.0001	.0004	.0007	.0016	.0046	.0109	.0178	.0223	.0410	.0692	.1094	1
6	.0000	.0000	.0000	.0000	.0000	.0000	.0000	.0000	.0000	.0000	.0000	.0000	.0001	.0002	.0007	.0014	.0018	.0041	.0083	.0156	0

n = 7

x	.01	.02	.03	.04	.05	.06	.07	.08	.09	.10	.15	1/6	.20	.25	.30	1/3	.35	.40	.45	.50	
0	1.000	1.000	1.000	1.000	1.000	1.000	1.000	1.000	1.000	1.000	1.000	1.000	1.000	1.000	1.000	1.000	1.000	1.000	1.000	1.000	7
1	.0679	.1319	.1920	.2486	.3017	.3515	.3983	.4422	.4832	.5217	.6794	.7209	.7903	.8665	.9176	.9415	.9510	.9720	.9848	.9922	6
2	.0020	.0079	.0171	.0294	.0444	.0618	.0813	.1026	.1255	.1497	.2834	.3302	.4233	.5551	.6706	.7366	.7662	.8414	.8976	.9375	5
3	.0000	.0003	.0009	.0020	.0038	.0063	.0097	.0140	.0193	.0257	.0738	.0958	.1480	.2436	.3529	.4294	.4677	.5801	.6836	.7734	4
4	.0000	.0000	.0000	.0001	.0002	.0004	.0007	.0012	.0018	.0027	.0121	.0176	.0333	.0706	.1260	.1733	.1998	.2898	.3917	.5000	3
5	.0000	.0000	.0000	.0000	.0000	.0000	.0000	.0001	.0001	.0002	.0012	.0020	.0047	.0129	.0288	.0453	.0556	.0963	.1529	.2266	2
6	.0000	.0000	.0000	.0000	.0000	.0000	.0000	.0000	.0000	.0000	.0001	.0001	.0004	.0013	.0038	.0069	.0090	.0188	.0357	.0625	1
7	.0000	.0000	.0000	.0000	.0000	.0000	.0000	.0000	.0000	.0000	.0000	.0000	.0000	.0001	.0002	.0005	.0006	.0016	.0037	.0078	0

n = 8

x	.01	.02	.03	.04	.05	.06	.07	.08	.09	.10	.15	1/6	.20	.25	.30	1/3	.35	.40	.45	.50	
0	1.000	1.000	1.000	1.000	1.000	1.000	1.000	1.000	1.000	1.000	1.000	1.000	1.000	1.000	1.000	1.000	1.000	1.000	1.000	1.000	8
1	.0773	.1492	.2163	.2786	.3366	.3904	.4404	.4868	.5297	.5695	.7275	.7674	.8322	.8999	.9424	.9610	.9681	.9832	.9916	.9961	7
2	.0027	.0103	.0223	.0381	.0572	.0792	.1035	.1298	.1577	.1869	.3428	.3953	.4967	.6329	.7447	.8049	.8309	.8936	.9368	.9648	6
3	.0001	.0004	.0013	.0031	.0058	.0096	.0147	.0211	.0289	.0381	.1052	.1348	.2031	.3215	.4482	.5318	.5722	.6846	.7799	.8555	5
4	.0000	.0000	.0001	.0002	.0004	.0007	.0013	.0022	.0034	.0050	.0214	.0307	.0563	.1138	.1941	.2586	.2936	.4059	.5230	.6367	4
5	.0000	.0000	.0000	.0000	.0000	.0000	.0001	.0001	.0003	.0004	.0029	.0046	.0104	.0273	.0580	.0879	.1061	.1737	.2604	.3633	3
6	.0000	.0000	.0000	.0000	.0000	.0000	.0000	.0000	.0000	.0000	.0002	.0004	.0012	.0042	.0113	.0197	.0253	.0498	.0885	.1445	2
7	.0000	.0000	.0000	.0000	.0000	.0000	.0000	.0000	.0000	.0000	.0000	.0000	.0001	.0004	.0013	.0026	.0036	.0085	.0181	.0352	1
8	.0000	.0000	.0000	.0000	.0000	.0000	.0000	.0000	.0000	.0000	.0000	.0000	.0000	.0000	.0001	.0002	.0002	.0007	.0017	.0039	0

p	.99	.98	.97	.96	.95	.94	.93	.92	.91	.90	.85	5/6	.80	.75	.70	2/3	.65	.60	.55	.50	p

Prob $(X \leqslant x)$

Pages 5, 7, 9 and 11 give cumulative probabilities for the same range of binomial distributions as covered on pages 4, 6, 8 and 10. For values of $p \leqslant \tfrac{1}{2}$ (along the top horizontal) refer to the x values in the left-hand column, the table entries giving Prob $(X \geqslant x)$; for values of $p \geqslant \tfrac{1}{2}$ (along the bottom horizontal) refer to the x-values in the right-hand column, the table entries giving Prob $(X \leqslant x)$ for these cases. Note that cumulative probabilities of the opposite type to those given may be calculated by Prob $(X \leqslant x) = 1 - \text{Prob}\,(X \geqslant x + 1)$ and Prob $(X \geqslant x) = 1 - \text{Prob}\,(X \leqslant x - 1)$.

The binomial distribution: individual probabilities

Prob $(X = x)$

p	.01	.02	.03	.04	.05	.06	.07	.08	.09	.10	.15	$\frac{1}{6}$	.20	.25	.30	$\frac{1}{3}$	.35	.40	.45	.50	x

$n = 9$

x	.01	.02	.03	.04	.05	.06	.07	.08	.09	.10	.15	$\frac{1}{6}$	.20	.25	.30	$\frac{1}{3}$	.35	.40	.45	.50	
0	.9135	.8337	.7602	.6925	.6302	.5730	.5204	.4722	.4279	.3874	.2316	.1938	.1342	.0751	.0404	.0260	.0207	.0101	.0046	.0020	9
1	.0830	.1531	.2116	.2597	.2985	.3292	.3525	.3695	.3809	.3874	.3679	.3489	.3020	.2253	.1556	.1171	.1004	.0605	.0339	.0176	8
2	.0034	.0125	.0262	.0433	.0629	.0840	.1061	.1285	.1507	.1722	.2597	.2791	.3020	.3003	.2668	.2341	.2162	.1612	.1110	.0703	7
3	.0001	.0006	.0019	.0042	.0077	.0125	.0186	.0261	.0348	.0446	.1069	.1302	.1762	.2336	.2668	.2731	.2716	.2508	.2119	.1641	6
4	.0000	.0000	.0001	.0003	.0006	.0012	.0021	.0034	.0052	.0074	.0283	.0391	.0661	.1168	.1715	.2048	.2194	.2508	.2600	.2461	5
5	.0000	.0000	.0000	.0000	.0000	.0001	.0002	.0003	.0005	.0008	.0050	.0078	.0165	.0389	.0735	.1024	.1181	.1672	.2128	.2461	4
6	.0000	.0000	.0000	.0000	.0000	.0000	.0000	.0000	.0000	.0001	.0006	.0010	.0028	.0087	.0210	.0341	.0424	.0743	.1160	.1641	3
7	.0000	.0000	.0000	.0000	.0000	.0000	.0000	.0000	.0000	.0000	.0000	.0001	.0003	.0012	.0039	.0073	.0098	.0212	.0407	.0703	2
8	.0000	.0000	.0000	.0000	.0000	.0000	.0000	.0000	.0000	.0000	.0000	.0000	.0000	.0001	.0004	.0009	.0013	.0035	.0083	.0176	1
9	.0000	.0000	.0000	.0000	.0000	.0000	.0000	.0000	.0000	.0000	.0000	.0000	.0000	.0000	.0000	.0001	.0001	.0003	.0008	.0020	0

$n = 10$

x	.01	.02	.03	.04	.05	.06	.07	.08	.09	.10	.15	$\frac{1}{6}$	.20	.25	.30	$\frac{1}{3}$	.35	.40	.45	.50	
0	.9044	.8171	.7374	.6648	.5987	.5386	.4840	.4344	.3894	.3487	.1969	.1615	.1074	.0563	.0282	.0173	.0135	.0060	.0025	.0010	10
1	.0914	.1667	.2281	.2770	.3151	.3438	.3643	.3777	.3851	.3874	.3474	.3230	.2684	.1877	.1211	.0867	.0725	.0403	.0207	.0098	9
2	.0042	.0153	.0317	.0519	.0746	.0988	.1234	.1478	.1714	.1937	.2759	.2907	.3020	.2816	.2335	.1951	.1757	.1209	.0763	.0439	8
3	.0001	.0008	.0026	.0058	.0105	.0168	.0248	.0343	.0452	.0574	.1298	.1550	.2013	.2503	.2668	.2601	.2522	.2150	.1665	.1172	7
4	.0000	.0000	.0001	.0004	.0010	.0019	.0033	.0052	.0078	.0112	.0401	.0543	.0881	.1460	.2001	.2276	.2377	.2508	.2384	.2051	6
5	.0000	.0000	.0000	.0000	.0001	.0001	.0003	.0005	.0009	.0015	.0085	.0130	.0264	.0584	.1029	.1366	.1536	.2007	.2340	.2461	5
6	.0000	.0000	.0000	.0000	.0000	.0000	.0000	.0000	.0001	.0001	.0012	.0022	.0055	.0162	.0368	.0569	.0689	.1115	.1596	.2051	4
7	.0000	.0000	.0000	.0000	.0000	.0000	.0000	.0000	.0000	.0000	.0001	.0002	.0008	.0031	.0090	.0163	.0212	.0425	.0746	.1172	3
8	.0000	.0000	.0000	.0000	.0000	.0000	.0000	.0000	.0000	.0000	.0000	.0000	.0001	.0004	.0014	.0030	.0043	.0106	.0229	.0439	2
9	.0000	.0000	.0000	.0000	.0000	.0000	.0000	.0000	.0000	.0000	.0000	.0000	.0000	.0000	.0001	.0003	.0005	.0016	.0042	.0098	1
10	.0000	.0000	.0000	.0000	.0000	.0000	.0000	.0000	.0000	.0000	.0000	.0000	.0000	.0000	.0000	.0000	.0000	.0001	.0003	.0010	0

$n = 11$

x	.01	.02	.03	.04	.05	.06	.07	.08	.09	.10	.15	$\frac{1}{6}$	.20	.25	.30	$\frac{1}{3}$	.35	.40	.45	.50	
0	.8953	.8007	.7153	.6382	.5688	.5063	.4501	.3996	.3544	.3138	.1673	.1346	.0859	.0422	.0198	.0116	.0088	.0036	.0014	.0005	11
1	.0995	.1798	.2433	.2925	.3293	.3555	.3727	.3823	.3855	.3835	.3248	.2961	.2362	.1549	.0932	.0636	.0518	.0266	.0125	.0054	10
2	.0050	.0183	.0376	.0609	.0867	.1135	.1403	.1662	.1906	.2131	.2866	.2961	.2953	.2581	.1998	.1590	.1395	.0887	.0513	.0269	9
3	.0002	.0011	.0035	.0076	.0137	.0217	.0317	.0434	.0566	.0710	.1517	.1777	.2215	.2581	.2568	.2384	.2254	.1774	.1259	.0806	8
4	.0000	.0000	.0002	.0006	.0014	.0028	.0048	.0075	.0112	.0158	.0536	.0711	.1107	.1721	.2201	.2384	.2428	.2365	.2060	.1611	7
5	.0000	.0000	.0000	.0000	.0001	.0002	.0005	.0009	.0015	.0025	.0132	.0199	.0388	.0803	.1321	.1669	.1830	.2207	.2360	.2256	6
6	.0000	.0000	.0000	.0000	.0000	.0000	.0000	.0001	.0002	.0003	.0023	.0040	.0097	.0268	.0566	.0835	.0985	.1471	.1931	.2256	5
7	.0000	.0000	.0000	.0000	.0000	.0000	.0000	.0000	.0000	.0000	.0003	.0006	.0017	.0064	.0173	.0298	.0379	.0701	.1128	.1611	4
8	.0000	.0000	.0000	.0000	.0000	.0000	.0000	.0000	.0000	.0000	.0000	.0001	.0002	.0011	.0037	.0075	.0102	.0234	.0462	.0806	3
9	.0000	.0000	.0000	.0000	.0000	.0000	.0000	.0000	.0000	.0000	.0000	.0000	.0000	.0001	.0005	.0012	.0018	.0052	.0126	.0269	2
10	.0000	.0000	.0000	.0000	.0000	.0000	.0000	.0000	.0000	.0000	.0000	.0000	.0000	.0000	.0000	.0001	.0002	.0007	.0021	.0054	1
11	.0000	.0000	.0000	.0000	.0000	.0000	.0000	.0000	.0000	.0000	.0000	.0000	.0000	.0000	.0000	.0000	.0000	.0000	.0002	.0005	0

$n = 12$

x	.01	.02	.03	.04	.05	.06	.07	.08	.09	.10	.15	$\frac{1}{6}$	.20	.25	.30	$\frac{1}{3}$	.35	.40	.45	.50	
0	.8864	.7847	.6938	.6127	.5404	.4759	.4186	.3677	.3225	.2824	.1422	.1122	.0687	.0317	.0138	.0077	.0057	.0022	.0008	.0002	12
1	.1074	.1922	.2575	.3064	.3413	.3645	.3781	.3837	.3827	.3766	.3012	.2692	.2062	.1267	.0712	.0462	.0368	.0174	.0075	.0029	11
2	.0060	.0216	.0438	.0702	.0988	.1280	.1565	.1835	.2082	.2301	.2924	.2961	.2835	.2323	.1678	.1272	.1088	.0639	.0339	.0161	10
3	.0002	.0015	.0045	.0098	.0173	.0272	.0393	.0532	.0686	.0852	.1720	.1974	.2362	.2581	.2397	.2120	.1954	.1419	.0923	.0537	9
4	.0000	.0001	.0003	.0009	.0021	.0039	.0067	.0104	.0153	.0213	.0683	.0888	.1329	.1936	.2311	.2384	.2367	.2128	.1700	.1208	8
5	.0000	.0000	.0000	.0001	.0002	.0004	.0008	.0014	.0024	.0038	.0193	.0284	.0532	.1032	.1585	.1908	.2039	.2270	.2225	.1934	7
6	.0000	.0000	.0000	.0000	.0000	.0000	.0001	.0001	.0003	.0005	.0040	.0066	.0155	.0401	.0792	.1113	.1281	.1766	.2124	.2256	6
7	.0000	.0000	.0000	.0000	.0000	.0000	.0000	.0000	.0000	.0000	.0006	.0011	.0033	.0115	.0291	.0477	.0591	.1009	.1489	.1934	5
8	.0000	.0000	.0000	.0000	.0000	.0000	.0000	.0000	.0000	.0000	.0001	.0001	.0005	.0024	.0078	.0149	.0199	.0420	.0762	.1208	4
9	.0000	.0000	.0000	.0000	.0000	.0000	.0000	.0000	.0000	.0000	.0000	.0000	.0001	.0004	.0015	.0033	.0048	.0125	.0277	.0537	3
10	.0000	.0000	.0000	.0000	.0000	.0000	.0000	.0000	.0000	.0000	.0000	.0000	.0000	.0000	.0002	.0005	.0008	.0025	.0068	.0161	2
11	.0000	.0000	.0000	.0000	.0000	.0000	.0000	.0000	.0000	.0000	.0000	.0000	.0000	.0000	.0000	.0000	.0001	.0003	.0010	.0029	1
12	.0000	.0000	.0000	.0000	.0000	.0000	.0000	.0000	.0000	.0000	.0000	.0000	.0000	.0000	.0000	.0000	.0000	.0000	.0001	.0002	0

$n = 13$

x	.01	.02	.03	.04	.05	.06	.07	.08	.09	.10	.15	$\frac{1}{6}$	.20	.25	.30	$\frac{1}{3}$	.35	.40	.45	.50	
0	.8775	.7690	.6730	.5882	.5133	.4474	.3893	.3383	.2935	.2542	.1209	.0935	.0550	.0238	.0097	.0051	.0037	.0013	.0004	.0001	13
1	.1152	.2040	.2706	.3186	.3512	.3712	.3809	.3824	.3773	.3672	.2774	.2430	.1787	.1029	.0540	.0334	.0259	.0113	.0045	.0016	12
2	.0070	.0250	.0502	.0797	.1109	.1422	.1720	.1995	.2239	.2448	.2937	.2916	.2680	.2059	.1388	.1002	.0836	.0453	.0220	.0095	11
3	.0003	.0019	.0057	.0122	.0214	.0333	.0475	.0636	.0812	.0997	.1900	.2138	.2457	.2517	.2181	.1837	.1651	.1107	.0660	.0349	10
4	.0000	.0001	.0004	.0013	.0028	.0053	.0089	.0138	.0201	.0277	.0838	.1069	.1535	.2097	.2337	.2296	.2222	.1845	.1350	.0873	9
5	.0000	.0000	.0000	.0001	.0003	.0006	.0012	.0022	.0036	.0055	.0266	.0385	.0691	.1258	.1803	.2067	.2154	.2214	.1989	.1571	8
6	.0000	.0000	.0000	.0000	.0000	.0001	.0001	.0003	.0005	.0008	.0063	.0103	.0230	.0559	.1030	.1378	.1546	.1968	.2169	.2095	7
7	.0000	.0000	.0000	.0000	.0000	.0000	.0000	.0000	.0000	.0001	.0011	.0021	.0058	.0186	.0442	.0689	.0833	.1312	.1775	.2095	6
8	.0000	.0000	.0000	.0000	.0000	.0000	.0000	.0000	.0000	.0000	.0001	.0003	.0011	.0047	.0142	.0258	.0336	.0656	.1089	.1571	5
9	.0000	.0000	.0000	.0000	.0000	.0000	.0000	.0000	.0000	.0000	.0000	.0000	.0001	.0009	.0034	.0072	.0101	.0243	.0495	.0873	4
10	.0000	.0000	.0000	.0000	.0000	.0000	.0000	.0000	.0000	.0000	.0000	.0000	.0000	.0001	.0006	.0014	.0022	.0065	.0162	.0349	3
11	.0000	.0000	.0000	.0000	.0000	.0000	.0000	.0000	.0000	.0000	.0000	.0000	.0000	.0000	.0001	.0002	.0003	.0012	.0036	.0095	2
12	.0000	.0000	.0000	.0000	.0000	.0000	.0000	.0000	.0000	.0000	.0000	.0000	.0000	.0000	.0000	.0000	.0000	.0001	.0005	.0016	1
13	.0000	.0000	.0000	.0000	.0000	.0000	.0000	.0000	.0000	.0000	.0000	.0000	.0000	.0000	.0000	.0000	.0000	.0000	.0000	.0001	0

.99	.98	.97	.96	.95	.94	.93	.92	.91	.90	.85	$\frac{5}{6}$	.80	.75	.70	$\frac{2}{3}$	.65	.60	.55	.50	p	x

Prob $(X = x)$

EXAMPLES: If ten dice are thrown, what is the probability of obtaining exactly two sixes? With $n = 10$ and $p = \frac{1}{6}$, Prob $(X = 2)$ is found from the table to be 0.2907.

If a treatment has a 90% success-rate, what is the probability that all of twelve treated patients recover? With $n = 12$ and $p = 0.9$, the table gives Prob $(X = 12) = 0.2824$.

The binomial distribution: cumulative probabilities

Prob $(X \geqslant x)$

x	.01	.02	.03	.04	.05	.06	.07	.08	.09	.10	.15	⅙	.20	.25	.30	⅓	.35	.40	.45	.50	x
n = 9																					
0	1.000	1.000	1.000	1.000	1.000	1.000	1.000	1.000	1.000	1.000	1.000	1.000	1.000	1.000	1.000	1.000	1.000	1.000	1.000	1.000	9
1	.0865	.1663	.2398	.3075	.3698	.4270	.4796	.5278	.5721	.6126	.7684	.8062	.8658	.9249	.9596	.9740	.9793	.9899	.9954	.9980	8
2	.0034	.0131	.0282	.0478	.0712	.0978	.1271	.1583	.1912	.2252	.4005	.4573	.5638	.6997	.8040	.8569	.8789	.9295	.9615	.9805	7
3	.0001	.0006	.0020	.0045	.0084	.0138	.0209	.0298	.0405	.0530	.1409	.1783	.2618	.3993	.5372	.6228	.6627	.7682	.8505	.9102	6
4	.0000	.0000	.0001	.0003	.0006	.0013	.0023	.0037	.0057	.0083	.0339	.0480	.0856	.1657	.2703	.3497	.3911	.5174	.6386	.7461	5
5	.0000	.0000	.0000	.0000	.0000	.0001	.0002	.0003	.0005	.0009	.0056	.0090	.0196	.0489	.0988	.1448	.1717	.2666	.3786	.5000	4
6	.0000	.0000	.0000	.0000	.0000	.0000	.0000	.0000	.0000	.0001	.0006	.0011	.0031	.0100	.0253	.0424	.0536	.0994	.1658	.2539	3
7	.0000	.0000	.0000	.0000	.0000	.0000	.0000	.0000	.0000	.0000	.0000	.0001	.0003	.0013	.0043	.0083	.0112	.0250	.0498	.0898	2
8	.0000	.0000	.0000	.0000	.0000	.0000	.0000	.0000	.0000	.0000	.0000	.0000	.0000	.0001	.0004	.0010	.0014	.0038	.0091	.0195	1
9	.0000	.0000	.0000	.0000	.0000	.0000	.0000	.0000	.0000	.0000	.0000	.0000	.0000	.0000	.0000	.0001	.0001	.0003	.0008	.0020	0
n = 10																					
0	1.000	1.000	1.000	1.000	1.000	1.000	1.000	1.000	1.000	1.000	1.000	1.000	1.000	1.000	1.000	1.000	1.000	1.000	1.000	1.000	10
1	.0956	.1829	.2626	.3352	.4013	.4614	.5160	.5656	.6106	.6513	.8031	.8385	.8926	.9437	.9718	.9827	.9865	.9940	.9975	.9990	9
2	.0043	.0162	.0345	.0582	.0861	.1176	.1517	.1879	.2254	.2639	.4557	.5155	.6242	.7560	.8507	.8960	.9140	.9536	.9767	.9893	8
3	.0001	.0009	.0028	.0062	.0115	.0188	.0283	.0401	.0540	.0702	.1798	.2248	.3222	.4744	.6172	.7009	.7384	.8327	.9004	.9453	7
4	.0000	.0000	.0001	.0004	.0010	.0020	.0036	.0058	.0088	.0128	.0500	.0697	.1209	.2241	.3504	.4407	.4862	.6177	.7340	.8281	6
5	.0000	.0000	.0000	.0000	.0001	.0002	.0003	.0006	.0010	.0016	.0099	.0155	.0328	.0781	.1503	.2131	.2485	.3669	.4956	.6230	5
6	.0000	.0000	.0000	.0000	.0000	.0000	.0000	.0000	.0001	.0001	.0014	.0024	.0064	.0197	.0473	.0766	.0949	.1662	.2616	.3770	4
7	.0000	.0000	.0000	.0000	.0000	.0000	.0000	.0000	.0000	.0000	.0001	.0003	.0009	.0035	.0106	.0197	.0260	.0548	.1020	.1719	3
8	.0000	.0000	.0000	.0000	.0000	.0000	.0000	.0000	.0000	.0000	.0000	.0000	.0001	.0004	.0016	.0034	.0048	.0123	.0274	.0547	2
9	.0000	.0000	.0000	.0000	.0000	.0000	.0000	.0000	.0000	.0000	.0000	.0000	.0000	.0000	.0001	.0004	.0005	.0017	.0045	.0107	1
10	.0000	.0000	.0000	.0000	.0000	.0000	.0000	.0000	.0000	.0000	.0000	.0000	.0000	.0000	.0000	.0000	.0000	.0001	.0003	.0010	0
n = 11																					
0	1.000	1.000	1.000	1.000	1.000	1.000	1.000	1.000	1.000	1.000	1.000	1.000	1.000	1.000	1.000	1.000	1.000	1.000	1.000	1.000	11
1	.1047	.1993	.2847	.3618	.4312	.4937	.5499	.6004	.6456	.6862	.8327	.8654	.9141	.9578	.9802	.9884	.9912	.9964	.9986	.9995	10
2	.0052	.0195	.0413	.0692	.1019	.1382	.1772	.2181	.2601	.3026	.5078	.5693	.6779	.8029	.8870	.9248	.9394	.9698	.9861	.9941	9
3	.0002	.0012	.0037	.0083	.0152	.0248	.0370	.0519	.0695	.0896	.2212	.2732	.3826	.5448	.6873	.7659	.7999	.8811	.9348	.9673	8
4	.0000	.0000	.0002	.0007	.0016	.0030	.0053	.0085	.0129	.0185	.0694	.0956	.1611	.2867	.4304	.5274	.5744	.7037	.8089	.8867	7
5	.0000	.0000	.0000	.0000	.0001	.0003	.0005	.0010	.0017	.0028	.0159	.0245	.0504	.1146	.2103	.2890	.3317	.4672	.6029	.7256	6
6	.0000	.0000	.0000	.0000	.0000	.0000	.0000	.0001	.0001	.0003	.0027	.0046	.0117	.0343	.0782	.1221	.1487	.2465	.3669	.5000	5
7	.0000	.0000	.0000	.0000	.0000	.0000	.0000	.0000	.0000	.0000	.0003	.0006	.0020	.0076	.0216	.0386	.0501	.0994	.1738	.2744	4
8	.0000	.0000	.0000	.0000	.0000	.0000	.0000	.0000	.0000	.0000	.0000	.0001	.0002	.0012	.0043	.0088	.0122	.0203	.0610	.1133	3
9	.0000	.0000	.0000	.0000	.0000	.0000	.0000	.0000	.0000	.0000	.0000	.0000	.0000	.0001	.0006	.0014	.0020	.0059	.0148	.0327	2
10	.0000	.0000	.0000	.0000	.0000	.0000	.0000	.0000	.0000	.0000	.0000	.0000	.0000	.0000	.0000	.0001	.0002	.0007	.0022	.0059	1
11	.0000	.0000	.0000	.0000	.0000	.0000	.0000	.0000	.0000	.0000	.0000	.0000	.0000	.0000	.0000	.0000	.0000	.0000	.0002	.0005	0
n = 12																					
0	1.000	1.000	1.000	1.000	1.000	1.000	1.000	1.000	1.000	1.000	1.000	1.000	1.000	1.000	1.000	1.000	1.000	1.000	1.000	1.000	12
1	.1136	.2153	.3062	.3873	.4596	.5241	.5814	.6323	.6775	.7176	.8578	.8878	.9313	.9683	.9862	.9923	.9943	.9978	.9992	.9998	11
2	.0062	.0231	.0486	.0809	.1184	.1595	.2033	.2487	.2948	.3410	.5565	.6187	.7251	.8416	.9150	.9460	.9576	.9804	.9917	.9968	10
3	.0002	.0015	.0048	.0107	.0196	.0316	.0468	.0652	.0866	.1109	.2642	.3226	.4417	.6093	.7472	.8189	.8487	.9166	.9579	.9807	9
4	.0000	.0001	.0003	.0010	.0022	.0043	.0075	.0120	.0180	.0256	.0922	.1252	.2054	.3512	.5075	.6069	.6533	.7747	.8655	.9270	8
5	.0000	.0000	.0000	.0001	.0002	.0004	.0009	.0016	.0027	.0043	.0239	.0364	.0726	.1576	.2763	.3685	.4167	.5618	.6956	.8062	7
6	.0000	.0000	.0000	.0000	.0000	.0000	.0001	.0002	.0003	.0005	.0046	.0079	.0194	.0544	.1178	.1777	.2127	.3348	.4731	.6128	6
7	.0000	.0000	.0000	.0000	.0000	.0000	.0000	.0000	.0000	.0001	.0007	.0013	.0039	.0143	.0386	.0664	.0846	.1582	.2607	.3872	5
8	.0000	.0000	.0000	.0000	.0000	.0000	.0000	.0000	.0000	.0000	.0001	.0002	.0006	.0028	.0095	.0188	.0255	.0573	.1117	.1938	4
9	.0000	.0000	.0000	.0000	.0000	.0000	.0000	.0000	.0000	.0000	.0000	.0000	.0001	.0004	.0017	.0039	.0056	.0153	.0356	.0730	3
10	.0000	.0000	.0000	.0000	.0000	.0000	.0000	.0000	.0000	.0000	.0000	.0000	.0000	.0000	.0002	.0005	.0008	.0028	.0079	.0193	2
11	.0000	.0000	.0000	.0000	.0000	.0000	.0000	.0000	.0000	.0000	.0000	.0000	.0000	.0000	.0000	.0000	.0001	.0003	.0011	.0032	1
12	.0000	.0000	.0000	.0000	.0000	.0000	.0000	.0000	.0000	.0000	.0000	.0000	.0000	.0000	.0000	.0000	.0000	.0000	.0001	.0002	0
n = 13																					
0	1.000	1.000	1.000	1.000	1.000	1.000	1.000	1.000	1.000	1.000	1.000	1.000	1.000	1.000	1.000	1.000	1.000	1.000	1.000	1.000	13
1	.1225	.2310	.3270	.4118	.4867	.5526	.6107	.6617	.7065	.7458	.8791	.9065	.9450	.9762	.9903	.9949	.9963	.9987	.9996	.9999	12
2	.0072	.0270	.0564	.0932	.1354	.1814	.2298	.2794	.3293	.3787	.6017	.6635	.7664	.8733	.9363	.9615	.9704	.9874	.9951	.9983	11
3	.0003	.0020	.0062	.0135	.0245	.0392	.0578	.0799	.1054	.1339	.3080	.3719	.4983	.6674	.7975	.8613	.8868	.9421	.9731	.9888	10
4	.0000	.0001	.0005	.0014	.0031	.0060	.0103	.0163	.0242	.0342	.1180	.1581	.2527	.4157	.5794	.6776	.7217	.8314	.9071	.9539	9
5	.0000	.0000	.0000	.0001	.0003	.0007	.0013	.0024	.0041	.0065	.0342	.0512	.0991	.2060	.3457	.4480	.4995	.6470	.7721	.8666	8
6	.0000	.0000	.0000	.0000	.0000	.0001	.0001	.0003	.0005	.0009	.0076	.0127	.0300	.0802	.1654	.2413	.2841	.4256	.5732	.7095	7
7	.0000	.0000	.0000	.0000	.0000	.0000	.0000	.0000	.0001	.0001	.0013	.0024	.0070	.0243	.0624	.1035	.1295	.2288	.3563	.5000	6
8	.0000	.0000	.0000	.0000	.0000	.0000	.0000	.0000	.0000	.0000	.0002	.0003	.0012	.0056	.0182	.0347	.0462	.0977	.1788	.2905	5
9	.0000	.0000	.0000	.0000	.0000	.0000	.0000	.0000	.0000	.0000	.0000	.0000	.0002	.0010	.0040	.0088	.0126	.0321	.0698	.1334	4
10	.0000	.0000	.0000	.0000	.0000	.0000	.0000	.0000	.0000	.0000	.0000	.0000	.0000	.0001	.0007	.0016	.0025	.0078	.0203	.0461	3
11	.0000	.0000	.0000	.0000	.0000	.0000	.0000	.0000	.0000	.0000	.0000	.0000	.0000	.0000	.0001	.0002	.0003	.0013	.0041	.0112	2
12	.0000	.0000	.0000	.0000	.0000	.0000	.0000	.0000	.0000	.0000	.0000	.0000	.0000	.0000	.0000	.0000	.0000	.0001	.0005	.0017	1
13	.0000	.0000	.0000	.0000	.0000	.0000	.0000	.0000	.0000	.0000	.0000	.0000	.0000	.0000	.0000	.0000	.0000	.0000	.0000	.0001	0
p	.99	.98	.97	.96	.95	.94	.93	.92	.91	.90	.85	⅚	.80	.75	.70	⅔	.65	.60	.55	.50	x

Prob $(X \leqslant x)$

EXAMPLES: If ten dice are thrown, what is the probability of obtaining at most two sixes? Now, Prob $(X \leqslant 2) = 1 -$ Prob $(X \geqslant 3)$. With $n = 10$ and $p = \frac{1}{6}$, the table gives Prob $(X \geqslant 3)$ as 0.2248, so Prob $(X \leqslant 2) = 1 - 0.2248 = 0.7752$. If a treatment has a 90% success-rate, what is the probability that no more than ten patients recover out of twelve who are treated? With $n = 12$ and $p = 0.9$, the table gives Prob $(X \leqslant 10) = 0.3410$.

The binomial distribution: individual probabilities

Prob $(X = x)$

$n = 14$

x \ p	.01	.02	.03	.04	.05	.06	.07	.08	.09	.10	.15	$\frac{1}{6}$	.20	.25	.30	$\frac{1}{3}$	.35	.40	.45	.50	
0	.8687	.7536	.6528	.5647	.4877	.4205	.3620	.3112	.2670	.2288	.1028	.0779	.0440	.0178	.0068	.0034	.0024	.0008	.0002	.0001	14
1	.1229	.2153	.2827	.3294	.3593	.3758	.3815	.3788	.3698	.3559	.2539	.2181	.1539	.0832	.0407	.0240	.0181	.0073	.0027	.0009	13
2	.0081	.0286	.0568	.0892	.1229	.1559	.1867	.2141	.2377	.2570	.2912	.2835	.2501	.1802	.1134	.0779	.0634	.0317	.0141	.0056	12
3	.0003	.0023	.0070	.0149	.0259	.0398	.0562	.0745	.0940	.1142	.2056	.2268	.2501	.2402	.1943	.1559	.1366	.0845	.0462	.0222	11
4	.0000	.0001	.0006	.0017	.0037	.0070	.0116	.0178	.0256	.0349	.0998	.1247	.1720	.2202	.2290	.2143	.2022	.1549	.1040	.0611	10
5	.0000	.0000	.0000	.0001	.0004	.0009	.0018	.0031	.0051	.0078	.0352	.0499	.0860	.1468	.1963	.2143	.2178	.2066	.1701	.1222	9
6	.0000	.0000	.0000	.0000	.0000	.0001	.0002	.0004	.0008	.0013	.0093	.0150	.0322	.0734	.1262	.1607	.1759	.2066	.2088	.1833	8
7	.0000	.0000	.0000	.0000	.0000	.0000	.0000	.0000	.0001	.0002	.0019	.0034	.0092	.0280	.0618	.0918	.1082	.1574	.1952	.2095	7
8	.0000	.0000	.0000	.0000	.0000	.0000	.0000	.0000	.0000	.0000	.0003	.0006	.0020	.0082	.0232	.0402	.0510	.0918	.1398	.1833	6
9	.0000	.0000	.0000	.0000	.0000	.0000	.0000	.0000	.0000	.0000	.0000	.0001	.0003	.0018	.0066	.0134	.0183	.0408	.0762	.1222	5
10	.0000	.0000	.0000	.0000	.0000	.0000	.0000	.0000	.0000	.0000	.0000	.0000	.0000	.0003	.0014	.0033	.0049	.0136	.0312	.0611	4
11	.0000	.0000	.0000	.0000	.0000	.0000	.0000	.0000	.0000	.0000	.0000	.0000	.0000	.0000	.0002	.0006	.0010	.0033	.0093	.0222	3
12	.0000	.0000	.0000	.0000	.0000	.0000	.0000	.0000	.0000	.0000	.0000	.0000	.0000	.0000	.0000	.0001	.0001	.0005	.0019	.0056	2
13	.0000	.0000	.0000	.0000	.0000	.0000	.0000	.0000	.0000	.0000	.0000	.0000	.0000	.0000	.0000	.0000	.0000	.0001	.0002	.0009	1
14	.0000	.0000	.0000	.0000	.0000	.0000	.0000	.0000	.0000	.0000	.0000	.0000	.0000	.0000	.0000	.0000	.0000	.0000	.0000	.0001	0

$n = 15$

x \ p	.01	.02	.03	.04	.05	.06	.07	.08	.09	.10	.15	$\frac{1}{6}$	.20	.25	.30	$\frac{1}{3}$	.35	.40	.45	.50	
0	.8601	.7386	.6333	.5421	.4633	.3953	.3367	.2863	.2430	.2059	.0874	.0649	.0352	.0134	.0047	.0023	.0016	.0005	.0001	.0000	15
1	.1303	.2261	.2938	.3388	.3658	.3785	.3801	.3734	.3605	.3432	.2312	.1947	.1319	.0668	.0305	.0171	.0126	.0047	.0016	.0005	14
2	.0092	.0323	.0636	.0988	.1348	.1691	.2003	.2273	.2496	.2669	.2856	.2726	.2309	.1559	.0916	.0599	.0476	.0219	.0090	.0032	13
3	.0004	.0029	.0085	.0178	.0307	.0468	.0653	.0857	.1070	.1285	.2184	.2363	.2501	.2252	.1700	.1299	.1110	.0634	.0318	.0139	12
4	.0000	.0002	.0008	.0022	.0049	.0090	.0148	.0223	.0317	.0428	.1156	.1418	.1876	.2252	.2186	.1948	.1792	.1268	.0780	.0417	11
5	.0000	.0000	.0001	.0002	.0006	.0013	.0024	.0043	.0069	.0105	.0449	.0624	.1032	.1651	.2061	.2143	.2123	.1859	.1404	.0916	10
6	.0000	.0000	.0000	.0000	.0000	.0001	.0003	.0006	.0011	.0019	.0132	.0208	.0430	.0917	.1472	.1786	.1906	.2066	.1914	.1527	9
7	.0000	.0000	.0000	.0000	.0000	.0000	.0000	.0001	.0001	.0003	.0030	.0053	.0138	.0393	.0811	.1148	.1319	.1771	.2013	.1964	8
8	.0000	.0000	.0000	.0000	.0000	.0000	.0000	.0000	.0000	.0000	.0005	.0011	.0035	.0131	.0348	.0574	.0710	.1181	.1647	.1964	7
9	.0000	.0000	.0000	.0000	.0000	.0000	.0000	.0000	.0000	.0000	.0001	.0002	.0007	.0034	.0116	.0223	.0298	.0612	.1048	.1527	6
10	.0000	.0000	.0000	.0000	.0000	.0000	.0000	.0000	.0000	.0000	.0000	.0000	.0001	.0007	.0030	.0067	.0096	.0245	.0515	.0916	5
11	.0000	.0000	.0000	.0000	.0000	.0000	.0000	.0000	.0000	.0000	.0000	.0000	.0000	.0001	.0006	.0015	.0024	.0074	.0191	.0417	4
12	.0000	.0000	.0000	.0000	.0000	.0000	.0000	.0000	.0000	.0000	.0000	.0000	.0000	.0000	.0001	.0003	.0004	.0016	.0052	.0139	3
13	.0000	.0000	.0000	.0000	.0000	.0000	.0000	.0000	.0000	.0000	.0000	.0000	.0000	.0000	.0000	.0000	.0001	.0003	.0010	.0032	2
14	.0000	.0000	.0000	.0000	.0000	.0000	.0000	.0000	.0000	.0000	.0000	.0000	.0000	.0000	.0000	.0000	.0000	.0000	.0001	.0005	1
15	.0000	.0000	.0000	.0000	.0000	.0000	.0000	.0000	.0000	.0000	.0000	.0000	.0000	.0000	.0000	.0000	.0000	.0000	.0000	.0000	0

$n = 16$

x \ p	.01	.02	.03	.04	.05	.06	.07	.08	.09	.10	.15	$\frac{1}{6}$	.20	.25	.30	$\frac{1}{3}$	.35	.40	.45	.50	
0	.8515	.7238	.6143	.5204	.4401	.3716	.3131	.2634	.2211	.1853	.0743	.0541	.0281	.0100	.0033	.0015	.0010	.0003	.0001	.0000	16
1	.1376	.2363	.3040	.3469	.3706	.3795	.3771	.3665	.3499	.3294	.2097	.1731	.1126	.0535	.0228	.0122	.0087	.0030	.0009	.0002	15
2	.0104	.0362	.0705	.1084	.1463	.1817	.2129	.2390	.2596	.2745	.2775	.2596	.2111	.1336	.0732	.0457	.0353	.0150	.0056	.0018	14
3	.0005	.0034	.0102	.0211	.0359	.0541	.0748	.0970	.1198	.1423	.2285	.2423	.2463	.2079	.1465	.1066	.0888	.0468	.0215	.0085	13
4	.0000	.0002	.0010	.0029	.0061	.0112	.0183	.0274	.0385	.0514	.1311	.1575	.2001	.2252	.2040	.1732	.1553	.1014	.0572	.0278	12
5	.0000	.0000	.0001	.0003	.0008	.0017	.0033	.0057	.0091	.0137	.0555	.0756	.1201	.1802	.2099	.2078	.2008	.1623	.1123	.0667	11
6	.0000	.0000	.0000	.0000	.0001	.0002	.0005	.0009	.0017	.0028	.0180	.0277	.0550	.1101	.1649	.1905	.1982	.1983	.1684	.1222	10
7	.0000	.0000	.0000	.0000	.0000	.0000	.0000	.0001	.0002	.0004	.0045	.0079	.0197	.0524	.1010	.1361	.1524	.1889	.1969	.1746	9
8	.0000	.0000	.0000	.0000	.0000	.0000	.0000	.0000	.0000	.0001	.0009	.0018	.0055	.0197	.0487	.0765	.0923	.1417	.1812	.1964	8
9	.0000	.0000	.0000	.0000	.0000	.0000	.0000	.0000	.0000	.0000	.0001	.0003	.0012	.0058	.0185	.0340	.0442	.0840	.1318	.1746	7
10	.0000	.0000	.0000	.0000	.0000	.0000	.0000	.0000	.0000	.0000	.0000	.0000	.0002	.0014	.0056	.0119	.0167	.0392	.0755	.1222	6
11	.0000	.0000	.0000	.0000	.0000	.0000	.0000	.0000	.0000	.0000	.0000	.0000	.0000	.0002	.0013	.0032	.0049	.0142	.0337	.0667	5
12	.0000	.0000	.0000	.0000	.0000	.0000	.0000	.0000	.0000	.0000	.0000	.0000	.0000	.0000	.0002	.0007	.0011	.0040	.0115	.0278	4
13	.0000	.0000	.0000	.0000	.0000	.0000	.0000	.0000	.0000	.0000	.0000	.0000	.0000	.0000	.0000	.0001	.0002	.0008	.0029	.0085	3
14	.0000	.0000	.0000	.0000	.0000	.0000	.0000	.0000	.0000	.0000	.0000	.0000	.0000	.0000	.0000	.0000	.0000	.0001	.0005	.0018	2
15	.0000	.0000	.0000	.0000	.0000	.0000	.0000	.0000	.0000	.0000	.0000	.0000	.0000	.0000	.0000	.0000	.0000	.0000	.0001	.0002	1
16	.0000	.0000	.0000	.0000	.0000	.0000	.0000	.0000	.0000	.0000	.0000	.0000	.0000	.0000	.0000	.0000	.0000	.0000	.0000	.0000	0

$n = 17$

x \ p	.01	.02	.03	.04	.05	.06	.07	.08	.09	.10	.15	$\frac{1}{6}$	.20	.25	.30	$\frac{1}{3}$	.35	.40	.45	.50	
0	.8429	.7093	.5958	.4996	.4181	.3493	.2912	.2423	.2012	.1668	.0631	.0451	.0225	.0075	.0023	.0010	.0007	.0002	.0000	.0000	17
1	.1447	.2461	.3133	.3539	.3741	.3790	.3726	.3582	.3383	.3150	.1893	.1532	.0957	.0426	.0169	.0086	.0060	.0019	.0005	.0001	16
2	.0117	.0402	.0775	.1180	.1575	.1935	.2244	.2492	.2677	.2800	.2673	.2452	.1914	.1136	.0581	.0345	.0260	.0102	.0035	.0010	15
3	.0006	.0041	.0120	.0246	.0415	.0618	.0844	.1083	.1324	.1556	.2359	.2452	.2393	.1893	.1245	.0863	.0701	.0341	.0144	.0052	14
4	.0000	.0003	.0013	.0036	.0076	.0138	.0222	.0330	.0458	.0605	.1457	.1716	.2093	.2209	.1868	.1510	.1320	.0796	.0411	.0182	13
5	.0000	.0000	.0001	.0004	.0010	.0023	.0044	.0075	.0118	.0175	.0668	.0893	.1361	.1914	.2081	.1963	.1849	.1379	.0875	.0472	12
6	.0000	.0000	.0000	.0000	.0001	.0003	.0007	.0013	.0023	.0039	.0236	.0357	.0680	.1276	.1784	.1963	.1991	.1839	.1432	.0944	11
7	.0000	.0000	.0000	.0000	.0000	.0000	.0001	.0002	.0004	.0007	.0065	.0112	.0267	.0668	.1201	.1542	.1685	.1927	.1841	.1484	10
8	.0000	.0000	.0000	.0000	.0000	.0000	.0000	.0000	.0000	.0001	.0014	.0028	.0084	.0279	.0644	.0964	.1134	.1606	.1883	.1855	9
9	.0000	.0000	.0000	.0000	.0000	.0000	.0000	.0000	.0000	.0000	.0003	.0006	.0021	.0093	.0276	.0482	.0611	.1070	.1540	.1855	8
10	.0000	.0000	.0000	.0000	.0000	.0000	.0000	.0000	.0000	.0000	.0000	.0001	.0004	.0025	.0095	.0193	.0263	.0571	.1008	.1484	7
11	.0000	.0000	.0000	.0000	.0000	.0000	.0000	.0000	.0000	.0000	.0000	.0000	.0001	.0005	.0026	.0061	.0090	.0242	.0525	.0944	6
12	.0000	.0000	.0000	.0000	.0000	.0000	.0000	.0000	.0000	.0000	.0000	.0000	.0000	.0001	.0006	.0015	.0024	.0081	.0215	.0472	5
13	.0000	.0000	.0000	.0000	.0000	.0000	.0000	.0000	.0000	.0000	.0000	.0000	.0000	.0000	.0001	.0003	.0005	.0021	.0068	.0182	4
14	.0000	.0000	.0000	.0000	.0000	.0000	.0000	.0000	.0000	.0000	.0000	.0000	.0000	.0000	.0000	.0000	.0001	.0004	.0016	.0052	3
15	.0000	.0000	.0000	.0000	.0000	.0000	.0000	.0000	.0000	.0000	.0000	.0000	.0000	.0000	.0000	.0000	.0000	.0001	.0003	.0010	2
16	.0000	.0000	.0000	.0000	.0000	.0000	.0000	.0000	.0000	.0000	.0000	.0000	.0000	.0000	.0000	.0000	.0000	.0000	.0000	.0001	1
17	.0000	.0000	.0000	.0000	.0000	.0000	.0000	.0000	.0000	.0000	.0000	.0000	.0000	.0000	.0000	.0000	.0000	.0000	.0000	.0000	0
	.99	.98	.97	.96	.95	.94	.93	.92	.91	.90	.85	$\frac{5}{6}$	.80	.75	.70	$\frac{2}{3}$	.65	.60	.55	.50	p / x

Prob $(X = x)$

The binomial distribution: cumulative probabilities

Prob $(X \geqslant x)$

n = 14

x	.01	.02	.03	.04	.05	.06	.07	.08	.09	.10	.15	1/6	.20	.25	.30	1/3	.35	.40	.45	.50	x
0	1.000	1.000	1.000	1.000	1.000	1.000	1.000	1.000	1.000	1.000	1.000	1.000	1.000	1.000	1.000	1.000	1.000	1.000	1.000	1.000	14
1	.1313	.2464	.3472	.4353	.5123	.5795	.6380	.6888	.7330	.7712	.8972	.9221	.9560	.9822	.9932	.9966	.9976	.9992	.9998	.9999	13
2	.0084	.0310	.0645	.1059	.1530	.2037	.2564	.3100	.3632	.4154	.6433	.7040	.8021	.8990	.9525	.9726	.9795	.9919	.9971	.9991	12
3	.0003	.0025	.0077	.0167	.0301	.0478	.0698	.0958	.1255	.1584	.3521	.4205	.5519	.7189	.8392	.8947	.9161	.9602	.9830	.9935	11
4	.0000	.0001	.0006	.0019	.0042	.0080	.0136	.0214	.0315	.0441	.1465	.1937	.3018	.4787	.6448	.7388	.7795	.8757	.9368	.9713	10
5	.0000	.0000	.0000	.0002	.0004	.0010	.0020	.0035	.0059	.0092	.0467	.0690	.1298	.2585	.4158	.5245	.5773	.7207	.8328	.9102	9
6	.0000	.0000	.0000	.0000	.0000	.0001	.0002	.0004	.0008	.0015	.0115	.0191	.0439	.1117	.2195	.3102	.3595	.5141	.6627	.7880	8
7	.0000	.0000	.0000	.0000	.0000	.0000	.0000	.0000	.0001	.0002	.0022	.0041	.0116	.0383	.0933	.1495	.1836	.3075	.4539	.6047	7
8	.0000	.0000	.0000	.0000	.0000	.0000	.0000	.0000	.0000	.0000	.0003	.0007	.0024	.0103	.0315	.0576	.0753	.1501	.2586	.3953	6
9	.0000	.0000	.0000	.0000	.0000	.0000	.0000	.0000	.0000	.0000	.0000	.0001	.0004	.0022	.0083	.0174	.0243	.0583	.1189	.2120	5
10	.0000	.0000	.0000	.0000	.0000	.0000	.0000	.0000	.0000	.0000	.0000	.0000	.0000	.0003	.0017	.0040	.0060	.0175	.0426	.0898	4
11	.0000	.0000	.0000	.0000	.0000	.0000	.0000	.0000	.0000	.0000	.0000	.0000	.0000	.0000	.0002	.0007	.0011	.0039	.0114	.0287	3
12	.0000	.0000	.0000	.0000	.0000	.0000	.0000	.0000	.0000	.0000	.0000	.0000	.0000	.0000	.0000	.0001	.0001	.0006	.0022	.0065	2
13	.0000	.0000	.0000	.0000	.0000	.0000	.0000	.0000	.0000	.0000	.0000	.0000	.0000	.0000	.0000	.0000	.0000	.0001	.0003	.0009	1
14	.0000	.0000	.0000	.0000	.0000	.0000	.0000	.0000	.0000	.0000	.0000	.0000	.0000	.0000	.0000	.0000	.0000	.0000	.0000	.0001	0

n = 15

x	.01	.02	.03	.04	.05	.06	.07	.08	.09	.10	.15	1/6	.20	.25	.30	1/3	.35	.40	.45	.50	x
0	1.000	1.000	1.000	1.000	1.000	1.000	1.000	1.000	1.000	1.000	1.000	1.000	1.000	1.000	1.000	1.000	1.000	1.000	1.000	1.000	15
1	.1399	.2614	.3667	.4579	.5367	.6047	.6633	.7137	.7570	.7941	.9126	.9351	.9648	.9866	.9953	.9977	.9984	.9995	.9999	1.000	14
2	.0096	.0353	.0730	.1191	.1710	.2262	.2832	.3403	.3965	.4510	.6814	.7404	.8329	.9198	.9647	.9806	.9858	.9948	.9983	.9995	13
3	.0004	.0030	.0094	.0203	.0362	.0571	.0829	.1130	.1469	.1841	.3958	.4678	.6020	.7639	.8732	.9206	.9383	.9729	.9893	.9963	12
4	.0000	.0002	.0008	.0024	.0055	.0104	.0175	.0273	.0399	.0556	.1773	.2315	.3518	.5387	.7031	.7908	.8273	.9095	.9576	.9824	11
5	.0000	.0000	.0001	.0002	.0006	.0014	.0028	.0050	.0082	.0127	.0617	.0898	.1642	.3135	.4845	.5959	.6481	.7827	.8796	.9408	10
6	.0000	.0000	.0000	.0000	.0001	.0001	.0003	.0007	.0013	.0022	.0168	.0274	.0611	.1484	.2784	.3816	.4357	.5968	.7392	.8491	9
7	.0000	.0000	.0000	.0000	.0000	.0000	.0000	.0001	.0002	.0003	.0036	.0066	.0181	.0566	.1311	.2030	.2452	.3902	.5478	.6964	8
8	.0000	.0000	.0000	.0000	.0000	.0000	.0000	.0000	.0000	.0000	.0006	.0013	.0042	.0173	.0500	.0882	.1132	.2131	.3465	.5000	7
9	.0000	.0000	.0000	.0000	.0000	.0000	.0000	.0000	.0000	.0000	.0001	.0002	.0008	.0042	.0152	.0308	.0422	.0950	.1818	.3036	6
10	.0000	.0000	.0000	.0000	.0000	.0000	.0000	.0000	.0000	.0000	.0000	.0000	.0001	.0008	.0037	.0085	.0124	.0338	.0769	.1509	5
11	.0000	.0000	.0000	.0000	.0000	.0000	.0000	.0000	.0000	.0000	.0000	.0000	.0000	.0001	.0007	.0018	.0028	.0093	.0255	.0592	4
12	.0000	.0000	.0000	.0000	.0000	.0000	.0000	.0000	.0000	.0000	.0000	.0000	.0000	.0000	.0001	.0003	.0005	.0019	.0063	.0176	3
13	.0000	.0000	.0000	.0000	.0000	.0000	.0000	.0000	.0000	.0000	.0000	.0000	.0000	.0000	.0000	.0000	.0001	.0003	.0011	.0037	2
14	.0000	.0000	.0000	.0000	.0000	.0000	.0000	.0000	.0000	.0000	.0000	.0000	.0000	.0000	.0000	.0000	.0000	.0000	.0001	.0005	1
15	.0000	.0000	.0000	.0000	.0000	.0000	.0000	.0000	.0000	.0000	.0000	.0000	.0000	.0000	.0000	.0000	.0000	.0000	.0000	.0000	0

n = 16

x	.01	.02	.03	.04	.05	.06	.07	.08	.09	.10	.15	1/6	.20	.25	.30	1/3	.35	.40	.45	.50	x
0	1.000	1.000	1.000	1.000	1.000	1.000	1.000	1.000	1.000	1.000	1.000	1.000	1.000	1.000	1.000	1.000	1.000	1.000	1.000	1.000	16
1	.1485	.2762	.3857	.4796	.5599	.6284	.6869	.7366	.7789	.8147	.9257	.9459	.9719	.9900	.9967	.9985	.9990	.9997	.9999	1.000	15
2	.0109	.0399	.0818	.1327	.1892	.2489	.3098	.3701	.4289	.4853	.7161	.7728	.8593	.9365	.9739	.9863	.9902	.9967	.9990	.9997	14
3	.0005	.0037	.0113	.0242	.0429	.0673	.0969	.1311	.1694	.2108	.4386	.5132	.6482	.8029	.9006	.9406	.9549	.9817	.9934	.9979	13
4	.0000	.0002	.0011	.0032	.0070	.0132	.0221	.0342	.0496	.0684	.2101	.2709	.4019	.5950	.7541	.8341	.8661	.9349	.9719	.9894	12
5	.0000	.0000	.0001	.0003	.0009	.0019	.0038	.0068	.0111	.0170	.0791	.1134	.2018	.3698	.5501	.6609	.7108	.8334	.9147	.9616	11
6	.0000	.0000	.0000	.0000	.0001	.0002	.0005	.0010	.0019	.0033	.0235	.0378	.0817	.1897	.3402	.4531	.5100	.6712	.8024	.8949	10
7	.0000	.0000	.0000	.0000	.0000	.0000	.0001	.0001	.0003	.0005	.0056	.0101	.0267	.0796	.1753	.2626	.3119	.4728	.6340	.7728	9
8	.0000	.0000	.0000	.0000	.0000	.0000	.0000	.0000	.0000	.0001	.0011	.0021	.0070	.0271	.0744	.1265	.1594	.2839	.4371	.5982	8
9	.0000	.0000	.0000	.0000	.0000	.0000	.0000	.0000	.0000	.0000	.0002	.0004	.0015	.0075	.0257	.0500	.0671	.1423	.2559	.4018	7
10	.0000	.0000	.0000	.0000	.0000	.0000	.0000	.0000	.0000	.0000	.0000	.0000	.0002	.0016	.0071	.0168	.0229	.0583	.1241	.2272	6
11	.0000	.0000	.0000	.0000	.0000	.0000	.0000	.0000	.0000	.0000	.0000	.0000	.0000	.0003	.0016	.0040	.0062	.0191	.0486	.1051	5
12	.0000	.0000	.0000	.0000	.0000	.0000	.0000	.0000	.0000	.0000	.0000	.0000	.0000	.0000	.0003	.0008	.0013	.0049	.0149	.0384	4
13	.0000	.0000	.0000	.0000	.0000	.0000	.0000	.0000	.0000	.0000	.0000	.0000	.0000	.0000	.0000	.0001	.0002	.0009	.0035	.0106	3
14	.0000	.0000	.0000	.0000	.0000	.0000	.0000	.0000	.0000	.0000	.0000	.0000	.0000	.0000	.0000	.0000	.0000	.0001	.0006	.0021	2
15	.0000	.0000	.0000	.0000	.0000	.0000	.0000	.0000	.0000	.0000	.0000	.0000	.0000	.0000	.0000	.0000	.0000	.0000	.0001	.0003	1
16	.0000	.0000	.0000	.0000	.0000	.0000	.0000	.0000	.0000	.0000	.0000	.0000	.0000	.0000	.0000	.0000	.0000	.0000	.0000	.0000	0

n = 17

x	.01	.02	.03	.04	.05	.06	.07	.08	.09	.10	.15	1/6	.20	.25	.30	1/3	.35	.40	.45	.50	x
0	1.000	1.000	1.000	1.000	1.000	1.000	1.000	1.000	1.000	1.000	1.000	1.000	1.000	1.000	1.000	1.000	1.000	1.000	1.000	1.000	17
1	.1571	.2907	.4042	.5004	.5819	.6507	.7088	.7577	.7988	.8332	.9369	.9549	.9775	.9925	.9977	.9990	.9993	.9998	1.000	1.000	16
2	.0123	.0446	.0909	.1465	.2078	.2717	.3362	.3995	.4604	.5182	.7475	.8017	.8818	.9499	.9807	.9904	.9933	.9979	.9994	.9999	15
3	.0006	.0044	.0134	.0286	.0503	.0782	.1118	.1503	.1927	.2382	.4802	.5565	.6904	.8363	.9226	.9558	.9673	.9877	.9959	.9988	14
4	.0000	.0003	.0014	.0040	.0088	.0164	.0273	.0419	.0603	.0826	.2444	.3113	.4511	.6470	.7981	.8696	.8972	.9536	.9816	.9936	13
5	.0000	.0000	.0001	.0004	.0012	.0026	.0051	.0089	.0145	.0221	.0987	.1396	.2418	.4261	.6113	.7186	.7652	.8740	.9404	.9755	12
6	.0000	.0000	.0000	.0000	.0001	.0003	.0007	.0015	.0027	.0047	.0319	.0504	.1057	.2347	.4032	.5223	.5803	.7361	.8529	.9283	11
7	.0000	.0000	.0000	.0000	.0000	.0000	.0001	.0002	.0004	.0008	.0083	.0147	.0377	.1071	.2248	.3261	.3812	.5522	.7098	.8338	10
8	.0000	.0000	.0000	.0000	.0000	.0000	.0000	.0000	.0000	.0001	.0017	.0035	.0109	.0402	.1046	.1719	.2128	.3595	.5257	.6855	9
9	.0000	.0000	.0000	.0000	.0000	.0000	.0000	.0000	.0000	.0000	.0003	.0007	.0026	.0124	.0403	.0755	.0994	.1989	.3374	.5000	8
10	.0000	.0000	.0000	.0000	.0000	.0000	.0000	.0000	.0000	.0000	.0000	.0001	.0005	.0031	.0127	.0273	.0383	.0919	.1834	.3145	7
11	.0000	.0000	.0000	.0000	.0000	.0000	.0000	.0000	.0000	.0000	.0000	.0000	.0001	.0006	.0032	.0080	.0120	.0348	.0826	.1662	6
12	.0000	.0000	.0000	.0000	.0000	.0000	.0000	.0000	.0000	.0000	.0000	.0000	.0000	.0001	.0007	.0019	.0030	.0106	.0301	.0717	5
13	.0000	.0000	.0000	.0000	.0000	.0000	.0000	.0000	.0000	.0000	.0000	.0000	.0000	.0000	.0001	.0003	.0006	.0025	.0086	.0245	4
14	.0000	.0000	.0000	.0000	.0000	.0000	.0000	.0000	.0000	.0000	.0000	.0000	.0000	.0000	.0000	.0000	.0001	.0006	.0019	.0064	3
15	.0000	.0000	.0000	.0000	.0000	.0000	.0000	.0000	.0000	.0000	.0000	.0000	.0000	.0000	.0000	.0000	.0000	.0001	.0003	.0012	2
16	.0000	.0000	.0000	.0000	.0000	.0000	.0000	.0000	.0000	.0000	.0000	.0000	.0000	.0000	.0000	.0000	.0000	.0000	.0000	.0001	1
17	.0000	.0000	.0000	.0000	.0000	.0000	.0000	.0000	.0000	.0000	.0000	.0000	.0000	.0000	.0000	.0000	.0000	.0000	.0000	.0000	0

| p | .99 | .98 | .97 | .96 | .95 | .94 | .93 | .92 | .91 | .90 | .85 | 5/6 | .80 | .75 | .70 | 2/3 | .65 | .60 | .55 | .50 | p |

Prob $(X \leqslant x)$

The binomial distribution: individual probabilities

Prob $(X = x)$

p	.01	.02	.03	.04	.05	.06	.07	.08	.09	.10	.15	$\frac{1}{6}$	.20	.25	.30	$\frac{1}{3}$	.35	.40	.45	.50	
x										$n = 18$											
0	.8345	.6951	.5780	.4796	.3972	.3283	.2708	.2229	.1831	.1501	.0536	.0376	.0180	.0056	.0016	.0007	.0004	.0001	.0000	.0000	18
1	.1517	.2554	.3217	.3597	.3763	.3772	.3669	.3489	.3260	.3002	.1704	.1352	.0811	.0338	.0126	.0061	.0042	.0012	.0003	.0001	17
2	.0130	.0443	.0846	.1274	.1683	.2047	.2348	.2579	.2741	.2835	.2556	.2299	.1723	.0958	.0458	.0259	.0190	.0069	.0022	.0006	16
3	.0007	.0048	.0140	.0283	.0473	.0697	.0942	.1196	.1446	.1680	.2406	.2452	.2297	.1704	.1046	.0690	.0547	.0246	.0095	.0031	15
4	.0000	.0004	.0016	.0044	.0093	.0167	.0266	.0390	.0536	.0700	.1592	.1839	.2153	.2130	.1681	.1294	.1104	.0614	.0291	.0117	14
5	.0000	.0000	.0001	.0005	.0014	.0030	.0056	.0095	.0148	.0218	.0787	.1030	.1507	.1988	.2017	.1812	.1664	.1146	.0666	.0327	13
6	.0000	.0000	.0000	.0000	.0002	.0004	.0009	.0018	.0032	.0052	.0301	.0446	.0816	.1436	.1873	.1963	.1941	.1655	.1181	.0708	12
7	.0000	.0000	.0000	.0000	.0000	.0000	.0001	.0003	.0005	.0010	.0091	.0153	.0350	.0820	.1376	.1682	.1792	.1892	.1657	.1214	11
8	.0000	.0000	.0000	.0000	.0000	.0000	.0000	.0000	.0001	.0002	.0022	.0042	.0120	.0376	.0811	.1157	.1327	.1734	.1864	.1669	10
9	.0000	.0000	.0000	.0000	.0000	.0000	.0000	.0000	.0000	.0000	.0004	.0009	.0033	.0139	.0386	.0643	.0794	.1284	.1694	.1855	9
10	.0000	.0000	.0000	.0000	.0000	.0000	.0000	.0000	.0000	.0000	.0001	.0002	.0008	.0042	.0149	.0289	.0385	.0771	.1248	.1669	8
11	.0000	.0000	.0000	.0000	.0000	.0000	.0000	.0000	.0000	.0000	.0000	.0000	.0001	.0010	.0046	.0105	.0151	.0374	.0742	.1214	7
12	.0000	.0000	.0000	.0000	.0000	.0000	.0000	.0000	.0000	.0000	.0000	.0000	.0000	.0002	.0012	.0031	.0047	.0145	.0354	.0708	6
13	.0000	.0000	.0000	.0000	.0000	.0000	.0000	.0000	.0000	.0000	.0000	.0000	.0000	.0000	.0002	.0007	.0012	.0045	.0134	.0327	5
14	.0000	.0000	.0000	.0000	.0000	.0000	.0000	.0000	.0000	.0000	.0000	.0000	.0000	.0000	.0000	.0001	.0002	.0011	.0039	.0117	4
15	.0000	.0000	.0000	.0000	.0000	.0000	.0000	.0000	.0000	.0000	.0000	.0000	.0000	.0000	.0000	.0000	.0000	.0002	.0009	.0031	3
16	.0000	.0000	.0000	.0000	.0000	.0000	.0000	.0000	.0000	.0000	.0000	.0000	.0000	.0000	.0000	.0000	.0000	.0000	.0001	.0006	2
17	.0000	.0000	.0000	.0000	.0000	.0000	.0000	.0000	.0000	.0000	.0000	.0000	.0000	.0000	.0000	.0000	.0000	.0000	.0000	.0001	1
18	.0000	.0000	.0000	.0000	.0000	.0000	.0000	.0000	.0000	.0000	.0000	.0000	.0000	.0000	.0000	.0000	.0000	.0000	.0000	.0000	0
										$n = 19$											
0	.8262	.6812	.5606	.4604	.3774	.3086	.2519	.2051	.1666	.1351	.0456	.0313	.0144	.0042	.0011	.0005	.0003	.0001	.0000	.0000	19
1	.1586	.2642	.3294	.3645	.3774	.3743	.3602	.3389	.3131	.2852	.1529	.1189	.0685	.0268	.0093	.0043	.0029	.0008	.0002	.0000	18
2	.0144	.0485	.0917	.1367	.1787	.2150	.2440	.2652	.2787	.2852	.2428	.2141	.1540	.0803	.0358	.0193	.0138	.0046	.0013	.0003	17
3	.0008	.0056	.0161	.0323	.0533	.0778	.1041	.1307	.1562	.1796	.2428	.2426	.2182	.1517	.0869	.0546	.0422	.0175	.0062	.0018	16
4	.0000	.0005	.0020	.0054	.0112	.0199	.0313	.0455	.0618	.0798	.1714	.1941	.2182	.2023	.1491	.1093	.0909	.0467	.0203	.0074	15
5	.0000	.0000	.0002	.0007	.0018	.0038	.0071	.0119	.0183	.0266	.0907	.1165	.1636	.2023	.1916	.1639	.1468	.0933	.0497	.0222	14
6	.0000	.0000	.0000	.0001	.0002	.0006	.0012	.0024	.0042	.0069	.0374	.0544	.0955	.1574	.1916	.1912	.1844	.1451	.0949	.0518	13
7	.0000	.0000	.0000	.0000	.0000	.0001	.0002	.0004	.0008	.0014	.0122	.0202	.0443	.0974	.1525	.1776	.1844	.1797	.1443	.0961	12
8	.0000	.0000	.0000	.0000	.0000	.0000	.0000	.0001	.0001	.0002	.0032	.0061	.0166	.0487	.0981	.1332	.1489	.1797	.1771	.1442	11
9	.0000	.0000	.0000	.0000	.0000	.0000	.0000	.0000	.0000	.0000	.0007	.0015	.0051	.0198	.0514	.0814	.0980	.1464	.1771	.1762	10
10	.0000	.0000	.0000	.0000	.0000	.0000	.0000	.0000	.0000	.0000	.0001	.0003	.0013	.0066	.0220	.0407	.0528	.0976	.1449	.1762	9
11	.0000	.0000	.0000	.0000	.0000	.0000	.0000	.0000	.0000	.0000	.0000	.0000	.0003	.0018	.0077	.0166	.0233	.0532	.0970	.1442	8
12	.0000	.0000	.0000	.0000	.0000	.0000	.0000	.0000	.0000	.0000	.0000	.0000	.0000	.0004	.0022	.0055	.0083	.0237	.0529	.0961	7
13	.0000	.0000	.0000	.0000	.0000	.0000	.0000	.0000	.0000	.0000	.0000	.0000	.0000	.0001	.0005	.0015	.0024	.0085	.0233	.0518	6
14	.0000	.0000	.0000	.0000	.0000	.0000	.0000	.0000	.0000	.0000	.0000	.0000	.0000	.0000	.0001	.0003	.0006	.0024	.0082	.0222	5
15	.0000	.0000	.0000	.0000	.0000	.0000	.0000	.0000	.0000	.0000	.0000	.0000	.0000	.0000	.0000	.0001	.0001	.0005	.0022	.0074	4
16	.0000	.0000	.0000	.0000	.0000	.0000	.0000	.0000	.0000	.0000	.0000	.0000	.0000	.0000	.0000	.0000	.0000	.0001	.0005	.0018	3
17	.0000	.0000	.0000	.0000	.0000	.0000	.0000	.0000	.0000	.0000	.0000	.0000	.0000	.0000	.0000	.0000	.0000	.0000	.0001	.0003	2
18	.0000	.0000	.0000	.0000	.0000	.0000	.0000	.0000	.0000	.0000	.0000	.0000	.0000	.0000	.0000	.0000	.0000	.0000	.0000	.0000	1
19	.0000	.0000	.0000	.0000	.0000	.0000	.0000	.0000	.0000	.0000	.0000	.0000	.0000	.0000	.0000	.0000	.0000	.0000	.0000	.0000	0
										$n = 20$											
0	.8179	.6676	.5438	.4420	.3585	.2901	.2342	.1887	.1516	.1216	.0388	.0261	.0115	.0032	.0008	.0003	.0002	.0000	.0000	.0000	20
1	.1652	.2725	.3364	.3683	.3774	.3703	.3526	.3282	.3000	.2702	.1368	.1043	.0576	.0211	.0068	.0030	.0020	.0005	.0001	.0000	19
2	.0159	.0528	.0988	.1458	.1887	.2246	.2521	.2711	.2818	.2852	.2293	.1982	.1369	.0669	.0278	.0143	.0100	.0031	.0008	.0002	18
3	.0010	.0065	.0183	.0364	.0596	.0860	.1139	.1414	.1672	.1901	.2428	.2379	.2054	.1339	.0716	.0429	.0323	.0123	.0040	.0011	17
4	.0000	.0006	.0024	.0065	.0133	.0233	.0364	.0523	.0703	.0898	.1821	.2022	.2182	.1897	.1304	.0911	.0738	.0350	.0139	.0046	16
5	.0000	.0000	.0002	.0009	.0022	.0048	.0088	.0145	.0222	.0319	.1028	.1294	.1746	.2023	.1789	.1457	.1272	.0746	.0365	.0148	15
6	.0000	.0000	.0000	.0001	.0003	.0008	.0017	.0032	.0055	.0089	.0454	.0647	.1091	.1686	.1916	.1821	.1712	.1244	.0746	.0370	14
7	.0000	.0000	.0000	.0000	.0000	.0001	.0002	.0005	.0011	.0020	.0160	.0259	.0545	.1124	.1643	.1821	.1844	.1659	.1221	.0739	13
8	.0000	.0000	.0000	.0000	.0000	.0000	.0000	.0001	.0002	.0004	.0046	.0084	.0222	.0609	.1144	.1480	.1614	.1797	.1623	.1201	12
9	.0000	.0000	.0000	.0000	.0000	.0000	.0000	.0000	.0000	.0001	.0011	.0022	.0074	.0271	.0654	.0987	.1158	.1597	.1771	.1602	11
10	.0000	.0000	.0000	.0000	.0000	.0000	.0000	.0000	.0000	.0000	.0002	.0005	.0020	.0099	.0308	.0543	.0686	.1171	.1593	.1762	10
11	.0000	.0000	.0000	.0000	.0000	.0000	.0000	.0000	.0000	.0000	.0000	.0001	.0005	.0030	.0120	.0247	.0336	.0710	.1185	.1602	9
12	.0000	.0000	.0000	.0000	.0000	.0000	.0000	.0000	.0000	.0000	.0000	.0000	.0001	.0008	.0039	.0092	.0136	.0355	.0727	.1201	8
13	.0000	.0000	.0000	.0000	.0000	.0000	.0000	.0000	.0000	.0000	.0000	.0000	.0000	.0002	.0010	.0028	.0045	.0146	.0366	.0739	7
14	.0000	.0000	.0000	.0000	.0000	.0000	.0000	.0000	.0000	.0000	.0000	.0000	.0000	.0000	.0002	.0007	.0012	.0049	.0150	.0370	6
15	.0000	.0000	.0000	.0000	.0000	.0000	.0000	.0000	.0000	.0000	.0000	.0000	.0000	.0000	.0000	.0001	.0003	.0013	.0049	.0148	5
16	.0000	.0000	.0000	.0000	.0000	.0000	.0000	.0000	.0000	.0000	.0000	.0000	.0000	.0000	.0000	.0000	.0000	.0003	.0013	.0046	4
17	.0000	.0000	.0000	.0000	.0000	.0000	.0000	.0000	.0000	.0000	.0000	.0000	.0000	.0000	.0000	.0000	.0000	.0000	.0002	.0011	3
18	.0000	.0000	.0000	.0000	.0000	.0000	.0000	.0000	.0000	.0000	.0000	.0000	.0000	.0000	.0000	.0000	.0000	.0000	.0000	.0002	2
19	.0000	.0000	.0000	.0000	.0000	.0000	.0000	.0000	.0000	.0000	.0000	.0000	.0000	.0000	.0000	.0000	.0000	.0000	.0000	.0000	1
20	.0000	.0000	.0000	.0000	.0000	.0000	.0000	.0000	.0000	.0000	.0000	.0000	.0000	.0000	.0000	.0000	.0000	.0000	.0000	.0000	0
	.99	.98	.97	.96	.95	.94	.93	.92	.91	.90	.85	$\frac{5}{6}$	.80	.75	.70	$\frac{2}{3}$	.65	.60	.55	.50	p / x

Prob $(X = x)$

The four charts on pages 12 and 13 are for use in binomial sampling experiments, both to find confidence intervals for p and to produce critical regions for the sample fraction $f = X/n$ (see bottom of page 4 for notation) when testing a null hypothesis H_0: $p = p_0$. The charts produce (a) confidence intervals having $\gamma = 90\%, 95\%, 98\%$ and 99% confidence levels; (b) one-sided critical regions (for alternative hypotheses H_1 of the form $p < p_0$ or $p > p_0$) for tests with significance levels $\alpha_1 = 5\%, 2\frac{1}{2}\%, 1\%$ and $\frac{1}{2}\%$; and (c) two-sided critical regions (for H_1 of the form $p \neq p_0$) for tests with significance levels $\alpha_2 = 10\%, 5\%, 2\%$ and 1%. For confidence intervals, locate the sample fraction f on the horizontal axis, trace up to the two curves labelled with the appropriate sample size n, and read off the confidence limits on the vertical axis. For critical regions, locate the hypothesised value of p, p_0, on the vertical axis, trace across to the two curves labelled with the sample size n and read off critical values f_1 and/or f_2 on the horizontal axis. If $f_1 < f_2$ the one-sided critical region for $H_1: p < p_0$ is $f \leqslant f_1$, or if H_1 is $p > p_0$ it is $f \geqslant f_2$. A two-sided critical

The binomial distribution: cumulative probabilities

Prob $(X \geq x)$

n = 18

x	.01	.02	.03	.04	.05	.06	.07	.08	.09	.10	.15	⅙	.20	.25	.30	⅓	.35	.40	.45	.50	x
0	1.000	1.000	1.000	1.000	1.000	1.000	1.000	1.000	1.000	1.000	1.000	1.000	1.000	1.000	1.000	1.000	1.000	1.000	1.000	1.000	18
1	.1655	.3049	.4220	.5204	.6028	.6717	.7292	.7771	.8169	.8499	.9464	.9624	.9820	.9944	.9984	.9993	.9996	.9999	1.000	1.000	17
2	.0138	.0495	.1003	.1607	.2265	.2945	.3622	.4281	.4909	.5497	.7759	.8272	.9009	.9605	.9858	.9932	.9954	.9987	.9997	.9999	16
3	.0007	.0052	.0157	.0333	.0581	.0898	.1275	.1702	.2168	.2662	.5203	.5973	.7287	.8647	.9400	.9674	.9764	.9918	.9975	.9993	15
4	.0000	.0004	.0018	.0050	.0109	.0201	.0333	.0506	.0723	.0982	.2798	.3521	.4990	.6943	.8354	.8983	.9217	.9672	.9880	.9962	14
5	.0000	.0000	.0002	.0006	.0015	.0034	.0067	.0116	.0186	.0282	.1206	.1682	.2836	.4813	.6673	.7689	.8114	.9058	.9589	.9846	13
6	.0000	.0000	.0000	.0001	.0002	.0005	.0010	.0021	.0038	.0064	.0419	.0653	.1329	.2825	.4656	.5878	.6450	.7912	.8923	.9519	12
7	.0000	.0000	.0000	.0000	.0000	.0000	.0001	.0003	.0006	.0012	.0118	.0206	.0513	.1390	.2783	.3915	.4509	.6257	.7742	.8811	11
8	.0000	.0000	.0000	.0000	.0000	.0000	.0000	.0000	.0001	.0002	.0027	.0053	.0163	.0569	.1407	.2233	.2717	.4366	.6085	.7597	10
9	.0000	.0000	.0000	.0000	.0000	.0000	.0000	.0000	.0000	.0000	.0005	.0011	.0043	.0193	.0596	.1076	.1391	.2632	.4222	.5927	9
10	.0000	.0000	.0000	.0000	.0000	.0000	.0000	.0000	.0000	.0000	.0001	.0002	.0009	.0054	.0210	.0433	.0597	.1347	.2527	.4073	8
11	.0000	.0000	.0000	.0000	.0000	.0000	.0000	.0000	.0000	.0000	.0000	.0000	.0002	.0012	.0061	.0144	.0212	.0576	.1280	.2403	7
12	.0000	.0000	.0000	.0000	.0000	.0000	.0000	.0000	.0000	.0000	.0000	.0000	.0000	.0002	.0014	.0039	.0062	.0203	.0537	.1189	6
13	.0000	.0000	.0000	.0000	.0000	.0000	.0000	.0000	.0000	.0000	.0000	.0000	.0000	.0000	.0003	.0009	.0014	.0058	.0183	.0481	5
14	.0000	.0000	.0000	.0000	.0000	.0000	.0000	.0000	.0000	.0000	.0000	.0000	.0000	.0000	.0000	.0001	.0003	.0013	.0049	.0154	4
15	.0000	.0000	.0000	.0000	.0000	.0000	.0000	.0000	.0000	.0000	.0000	.0000	.0000	.0000	.0000	.0000	.0000	.0002	.0010	.0038	3
16	.0000	.0000	.0000	.0000	.0000	.0000	.0000	.0000	.0000	.0000	.0000	.0000	.0000	.0000	.0000	.0000	.0000	.0000	.0001	.0007	2
17	.0000	.0000	.0000	.0000	.0000	.0000	.0000	.0000	.0000	.0000	.0000	.0000	.0000	.0000	.0000	.0000	.0000	.0000	.0000	.0001	1
18	.0000	.0000	.0000	.0000	.0000	.0000	.0000	.0000	.0000	.0000	.0000	.0000	.0000	.0000	.0000	.0000	.0000	.0000	.0000	.0000	0

n = 19

x	.01	.02	.03	.04	.05	.06	.07	.08	.09	.10	.15	⅙	.20	.25	.30	⅓	.35	.40	.45	.50	x
0	1.000	1.000	1.000	1.000	1.000	1.000	1.000	1.000	1.000	1.000	1.000	1.000	1.000	1.000	1.000	1.000	1.000	1.000	1.000	1.000	19
1	.1738	.3188	.4394	.5396	.6226	.6914	.7481	.7949	.8331	.8649	.9544	.9687	.9856	.9958	.9989	.9995	.9997	.9999	1.000	1.000	18
2	.0153	.0546	.1100	.1761	.2453	.3171	.3879	.4560	.5202	.5797	.8015	.8498	.9171	.9690	.9896	.9953	.9969	.9992	.9998	1.000	17
3	.0009	.0061	.0183	.0384	.0665	.1021	.1439	.1908	.2415	.2946	.5587	.6357	.7631	.8887	.9538	.9760	.9830	.9945	.9985	.9996	16
4	.0000	.0005	.0022	.0061	.0132	.0243	.0398	.0602	.0853	.1150	.3159	.3930	.5449	.7369	.8668	.9213	.9409	.9770	.9923	.9978	15
5	.0000	.0000	.0002	.0007	.0020	.0044	.0086	.0147	.0235	.0352	.1444	.1989	.3267	.5346	.7178	.8121	.8500	.9304	.9720	.9904	14
6	.0000	.0000	.0000	.0001	.0002	.0006	.0014	.0029	.0051	.0086	.0537	.0824	.1631	.3322	.5261	.6481	.7032	.8371	.9223	.9682	13
7	.0000	.0000	.0000	.0000	.0000	.0001	.0002	.0004	.0009	.0017	.0163	.0281	.0676	.1749	.3345	.4569	.5188	.6919	.8273	.9165	12
8	.0000	.0000	.0000	.0000	.0000	.0000	.0000	.0001	.0001	.0003	.0041	.0079	.0233	.0775	.1820	.2793	.3344	.5122	.6831	.8204	11
9	.0000	.0000	.0000	.0000	.0000	.0000	.0000	.0000	.0000	.0000	.0008	.0018	.0067	.0287	.0839	.1462	.1855	.3325	.5060	.6762	10
10	.0000	.0000	.0000	.0000	.0000	.0000	.0000	.0000	.0000	.0000	.0001	.0004	.0016	.0089	.0326	.0648	.0875	.1861	.3290	.5000	9
11	.0000	.0000	.0000	.0000	.0000	.0000	.0000	.0000	.0000	.0000	.0000	.0001	.0003	.0023	.0105	.0241	.0347	.0885	.1841	.3238	8
12	.0000	.0000	.0000	.0000	.0000	.0000	.0000	.0000	.0000	.0000	.0000	.0000	.0000	.0005	.0028	.0074	.0114	.0352	.0871	.1796	7
13	.0000	.0000	.0000	.0000	.0000	.0000	.0000	.0000	.0000	.0000	.0000	.0000	.0000	.0001	.0006	.0019	.0031	.0116	.0342	.0835	6
14	.0000	.0000	.0000	.0000	.0000	.0000	.0000	.0000	.0000	.0000	.0000	.0000	.0000	.0000	.0001	.0004	.0007	.0031	.0109	.0318	5
15	.0000	.0000	.0000	.0000	.0000	.0000	.0000	.0000	.0000	.0000	.0000	.0000	.0000	.0000	.0000	.0001	.0001	.0006	.0028	.0096	4
16	.0000	.0000	.0000	.0000	.0000	.0000	.0000	.0000	.0000	.0000	.0000	.0000	.0000	.0000	.0000	.0000	.0000	.0001	.0005	.0022	3
17	.0000	.0000	.0000	.0000	.0000	.0000	.0000	.0000	.0000	.0000	.0000	.0000	.0000	.0000	.0000	.0000	.0000	.0000	.0001	.0004	2
18	.0000	.0000	.0000	.0000	.0000	.0000	.0000	.0000	.0000	.0000	.0000	.0000	.0000	.0000	.0000	.0000	.0000	.0000	.0000	.0000	1
19	.0000	.0000	.0000	.0000	.0000	.0000	.0000	.0000	.0000	.0000	.0000	.0000	.0000	.0000	.0000	.0000	.0000	.0000	.0000	.0000	0

n = 20

x	.01	.02	.03	.04	.05	.06	.07	.08	.09	.10	.15	⅙	.20	.25	.30	⅓	.35	.40	.45	.50	x
0	1.000	1.000	1.000	1.000	1.000	1.000	1.000	1.000	1.000	1.000	1.000	1.000	1.000	1.000	1.000	1.000	1.000	1.000	1.000	1.000	20
1	.1821	.3324	.4562	.5580	.6415	.7099	.7658	.8113	.8484	.8784	.9612	.9739	.9885	.9968	.9992	.9997	.9998	1.000	1.000	1.000	19
2	.0169	.0599	.1198	.1897	.2642	.3395	.4131	.4831	.5484	.6083	.8244	.8696	.9308	.9757	.9924	.9967	.9979	.9995	.9999	1.000	18
3	.0010	.0071	.0210	.0439	.0755	.1150	.1610	.2121	.2666	.3231	.5951	.6713	.7939	.9087	.9645	.9824	.9879	.9964	.9991	.9998	17
4	.0000	.0006	.0027	.0074	.0159	.0290	.0471	.0706	.0993	.1330	.3523	.4335	.5886	.7748	.8929	.9396	.9556	.9840	.9951	.9987	16
5	.0000	.0000	.0003	.0010	.0026	.0056	.0107	.0183	.0290	.0432	.1702	.2313	.3704	.5852	.7625	.8485	.8818	.9490	.9811	.9941	15
6	.0000	.0000	.0000	.0001	.0003	.0009	.0019	.0038	.0068	.0113	.0673	.1018	.1958	.3828	.5836	.7028	.7546	.8744	.9447	.9793	14
7	.0000	.0000	.0000	.0000	.0000	.0001	.0003	.0006	.0013	.0024	.0219	.0371	.0867	.2142	.3920	.5207	.5834	.7500	.8701	.9423	13
8	.0000	.0000	.0000	.0000	.0000	.0000	.0000	.0001	.0002	.0004	.0059	.0113	.0321	.1018	.2277	.3385	.3990	.5841	.7480	.8684	12
9	.0000	.0000	.0000	.0000	.0000	.0000	.0000	.0000	.0000	.0001	.0013	.0028	.0100	.0409	.1133	.1905	.2376	.4044	.5857	.7483	11
10	.0000	.0000	.0000	.0000	.0000	.0000	.0000	.0000	.0000	.0000	.0002	.0006	.0026	.0139	.0480	.0919	.1218	.2447	.4086	.5881	10
11	.0000	.0000	.0000	.0000	.0000	.0000	.0000	.0000	.0000	.0000	.0000	.0001	.0006	.0039	.0171	.0376	.0532	.1275	.2493	.4119	9
12	.0000	.0000	.0000	.0000	.0000	.0000	.0000	.0000	.0000	.0000	.0000	.0000	.0001	.0009	.0051	.0130	.0196	.0565	.1308	.2517	8
13	.0000	.0000	.0000	.0000	.0000	.0000	.0000	.0000	.0000	.0000	.0000	.0000	.0000	.0002	.0013	.0037	.0060	.0210	.0580	.1316	7
14	.0000	.0000	.0000	.0000	.0000	.0000	.0000	.0000	.0000	.0000	.0000	.0000	.0000	.0000	.0003	.0009	.0015	.0065	.0214	.0577	6
15	.0000	.0000	.0000	.0000	.0000	.0000	.0000	.0000	.0000	.0000	.0000	.0000	.0000	.0000	.0000	.0002	.0003	.0016	.0064	.0207	5
16	.0000	.0000	.0000	.0000	.0000	.0000	.0000	.0000	.0000	.0000	.0000	.0000	.0000	.0000	.0000	.0000	.0000	.0003	.0015	.0059	4
17	.0000	.0000	.0000	.0000	.0000	.0000	.0000	.0000	.0000	.0000	.0000	.0000	.0000	.0000	.0000	.0000	.0000	.0000	.0003	.0013	3
18	.0000	.0000	.0000	.0000	.0000	.0000	.0000	.0000	.0000	.0000	.0000	.0000	.0000	.0000	.0000	.0000	.0000	.0000	.0000	.0002	2
19	.0000	.0000	.0000	.0000	.0000	.0000	.0000	.0000	.0000	.0000	.0000	.0000	.0000	.0000	.0000	.0000	.0000	.0000	.0000	.0000	1
20	.0000	.0000	.0000	.0000	.0000	.0000	.0000	.0000	.0000	.0000	.0000	.0000	.0000	.0000	.0000	.0000	.0000	.0000	.0000	.0000	0

| | .99 | .98 | .97 | .96 | .95 | .94 | .93 | .92 | .91 | .90 | .85 | ⅚ | .80 | .75 | .70 | ⅔ | .65 | .60 | .55 | .50 | p |

Prob $(X \leq x)$

region appropriate for $H_1: p \neq p_0$ is comprised of both of these one-sided regions. The 'curves' are in fact drawn as straight lines joining points corresponding to all $n + 1$ possible values of f (this is seen most clearly for small n). Use of values of f_1 and f_2 which are in fact not realisable values of f result in conservative critical regions, i.e. actual α_1 or α_2 values which are less than the nominal values.

EXAMPLES: With eight successes out of twenty, i.e. $n = 20$, $X = 8$ and $f = 8/20 = 0.4$, the $\gamma = 95\%$ confidence interval for p is (0.19:0.64), using the second chart on page 12. Using the same chart, suppose we wish to test $H_0: p = 0.6$, again with $n = 20$. We read off $f_1 = 0.36$ and $f_2 = 0.83$. So $f \leq 0.36$ (i.e. $X \leq 7$) is the $\alpha_1^L = 2\frac{1}{2}\%$ critical region appropriate for H_1: $p < 0.6$, $f \geq 0.83$ (i.e. $X \geq 17$) is the $\alpha_1^R = 2\frac{1}{2}\%$ critical region appropriate for $H_1: p > 0.6$, and these two regions combined constitute the $\alpha_2 = 5\%$ critical region appropriate for $H_1: p \neq 0.6$. α_1^L and α_1^R denote significance levels for the one-sided tests where H_1 says that p is to the Left or Right respectively of p_0. The true significance levels here are in all cases slightly less than the nominal figures of $2\frac{1}{2}\%$ or 5%.

Charts giving confidence intervals for p and critical values for the sample fraction

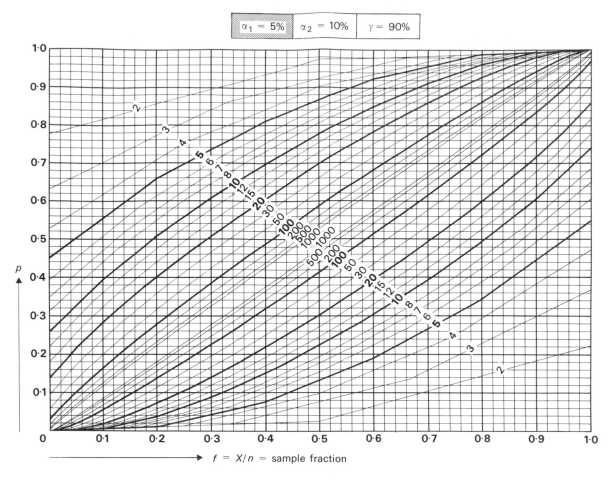

$\alpha_1 = 5\%$ | $\alpha_2 = 10\%$ | $\gamma = 90\%$

$f = X/n$ = sample fraction

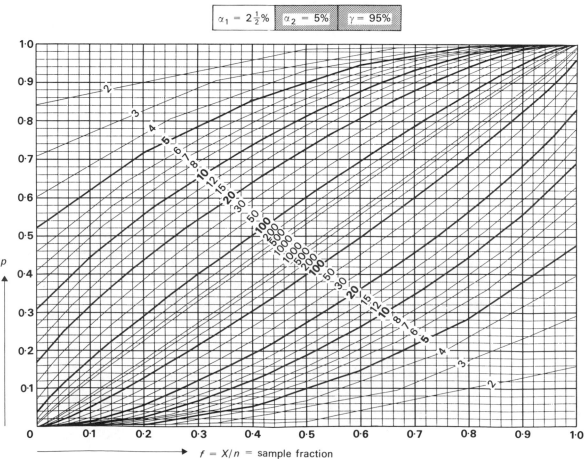

$\alpha_1 = 2\frac{1}{2}\%$ | $\alpha_2 = 5\%$ | $\gamma = 95\%$

$f = X/n$ = sample fraction

For description, see pages 10 and 11.

Charts giving confidence intervals for p and critical values for the sample fraction

$\alpha_1 = 1\%$	$\alpha_2 = 2\%$	$\gamma = 98\%$

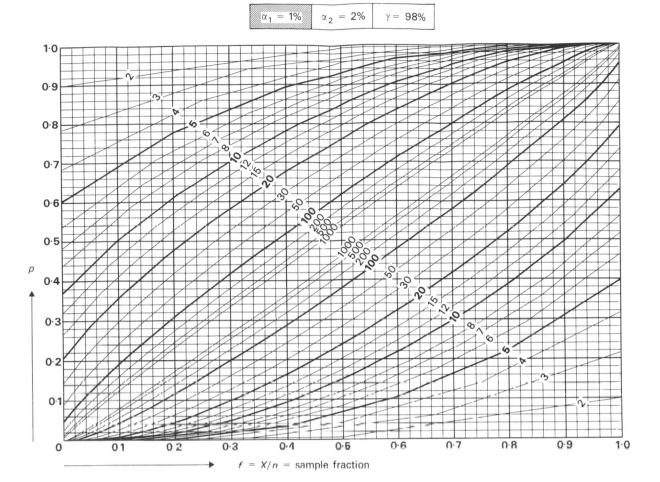

$f = X/n$ = sample fraction

$\alpha_1 = \frac{1}{2}\%$	$\alpha_2 = 1\%$	$\gamma = 99\%$

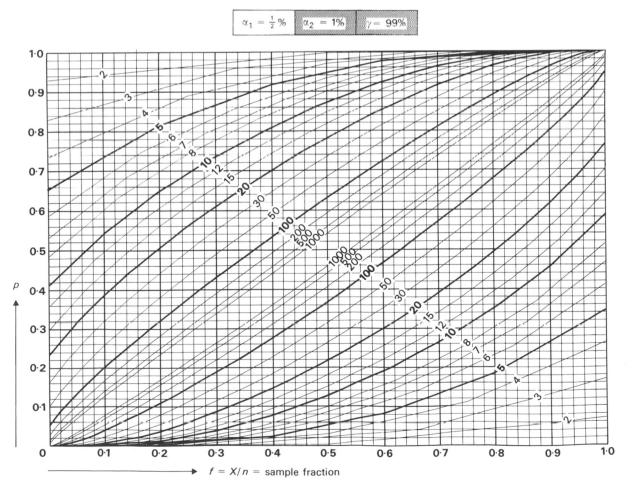

$f = X/n$ = sample fraction

For description, see pages 10 and 11.

The Poisson distribution: individual probabilities

$$\text{Prob}\,(X = x) \;=\; e^{-\mu}\cdot\frac{\mu^{x}}{x!}\qquad (x = 0, 1, 2, \ldots)$$

Prob $(X = x)$

μ	0.01	0.02	0.03	0.04	0.05	0.06	0.07	0.08	0.09	0.10	0.12	0.14	0.16	0.18	0.20	0.25	0.30	0.35	x
0	.9900	.9802	.9704	.9608	.9512	.9418	.9324	.9231	.9139	.9048	.8869	.8694	.8521	.8353	.8187	.7788	.7408	.7047	0
1	.0099	.0196	.0291	.0384	.0476	.0565	.0653	.0738	.0823	.0905	.1064	.1217	.1363	.1503	.1637	.1947	.2222	.2466	1
2	.0000	.0002	.0004	.0008	.0012	.0017	.0023	.0030	.0037	.0045	.0064	.0085	.0109	.0135	.0164	.0243	.0333	.0432	2
3	.0000	.0000	.0000	.0000	.0000	.0000	.0001	.0001	.0001	.0002	.0003	.0004	.0006	.0008	.0011	.0020	.0033	.0050	3
4	.0000	.0000	.0000	.0000	.0000	.0000	.0000	.0000	.0000	.0000	.0000	.0000	.0000	.0000	.0001	.0001	.0003	.0004	4
5	.0000	.0000	.0000	.0000	.0000	.0000	.0000	.0000	.0000	.0000	.0000	.0000	.0000	.0000	.0000	.0000	.0000	.0000	5

μ	0.40	0.45	0.50	0.55	0.60	0.65	0.70	0.75	0.80	0.85	0.90	0.95	1.00	1.10	1.20	1.30	1.40	1.50	x
0	.6703	.6376	.6065	.5769	.5488	.5220	.4966	.4724	.4493	.4274	.4066	.3867	.3679	.3329	.3012	.2725	.2466	.2231	0
1	.2681	.2869	.3033	.3173	.3293	.3393	.3476	.3543	.3595	.3633	.3659	.3674	.3679	.3662	.3614	.3543	.3452	.3347	1
2	.0536	.0646	.0758	.0873	.0988	.1103	.1217	.1329	.1438	.1544	.1647	.1745	.1839	.2014	.2169	.2303	.2417	.2510	2
3	.0072	.0097	.0126	.0160	.0198	.0239	.0284	.0332	.0383	.0437	.0494	.0553	.0613	.0738	.0867	.0998	.1128	.1255	3
4	.0007	.0011	.0016	.0022	.0030	.0039	.0050	.0062	.0077	.0093	.0111	.0131	.0153	.0203	.0260	.0324	.0395	.0471	4
5	.0001	.0001	.0002	.0002	.0004	.0005	.0007	.0009	.0012	.0016	.0020	.0025	.0031	.0045	.0062	.0084	.0111	.0141	5
6	.0000	.0000	.0000	.0000	.0000	.0001	.0001	.0001	.0002	.0002	.0003	.0004	.0005	.0008	.0012	.0018	.0026	.0035	6
7	.0000	.0000	.0000	.0000	.0000	.0000	.0000	.0000	.0000	.0000	.0000	.0001	.0001	.0001	.0002	.0003	.0005	.0008	7
8	.0000	.0000	.0000	.0000	.0000	.0000	.0000	.0000	.0000	.0000	.0000	.0000	.0000	.0000	.0000	.0001	.0001	.0001	8
9	.0000	.0000	.0000	.0000	.0000	.0000	.0000	.0000	.0000	.0000	.0000	.0000	.0000	.0000	.0000	.0000	.0000	.0000	9

μ	1.60	1.70	1.80	1.90	2.00	2.10	2.20	2.30	2.40	2.50	2.60	2.70	2.80	2.90	3.00	3.10	3.20	3.30	x
0	.2019	.1827	.1653	.1496	.1353	.1225	.1108	.1003	.0907	.0821	.0743	.0672	.0608	.0550	.0498	.0450	.0408	.0369	0
1	.3230	.3106	.2975	.2842	.2707	.2572	.2438	.2306	.2177	.2052	.1931	.1815	.1703	.1596	.1494	.1397	.1304	.1217	1
2	.2584	.2640	.2678	.2700	.2707	.2700	.2681	.2652	.2613	.2565	.2510	.2450	.2384	.2314	.2240	.2165	.2087	.2008	2
3	.1378	.1496	.1607	.1710	.1804	.1890	.1966	.2033	.2090	.2138	.2176	.2205	.2225	.2237	.2240	.2237	.2226	.2209	3
4	.0551	.0636	.0723	.0812	.0902	.0992	.1082	.1169	.1254	.1336	.1414	.1488	.1557	.1622	.1680	.1733	.1781	.1823	4
5	.0176	.0216	.0260	.0309	.0361	.0417	.0476	.0538	.0602	.0668	.0735	.0804	.0872	.0940	.1008	.1075	.1140	.1203	5
6	.0047	.0061	.0078	.0098	.0120	.0146	.0174	.0206	.0241	.0278	.0319	.0362	.0407	.0455	.0504	.0555	.0608	.0662	6
7	.0011	.0015	.0020	.0027	.0034	.0044	.0055	.0068	.0083	.0099	.0118	.0139	.0163	.0188	.0216	.0246	.0278	.0312	7
8	.0002	.0003	.0005	.0006	.0009	.0011	.0015	.0019	.0025	.0031	.0038	.0047	.0057	.0068	.0081	.0095	.0111	.0129	8
9	.0000	.0001	.0001	.0001	.0002	.0003	.0004	.0005	.0007	.0009	.0011	.0014	.0018	.0022	.0027	.0033	.0040	.0047	9
10	.0000	.0000	.0000	.0000	.0000	.0001	.0001	.0001	.0002	.0002	.0003	.0004	.0005	.0006	.0008	.0010	.0013	.0016	10
11	.0000	.0000	.0000	.0000	.0000	.0000	.0000	.0000	.0000	.0000	.0001	.0001	.0001	.0002	.0002	.0003	.0004	.0005	11
12	.0000	.0000	.0000	.0000	.0000	.0000	.0000	.0000	.0000	.0000	.0000	.0000	.0000	.0000	.0001	.0001	.0001	.0001	12
13	.0000	.0000	.0000	.0000	.0000	.0000	.0000	.0000	.0000	.0000	.0000	.0000	.0000	.0000	.0000	.0000	.0000	.0000	13

μ	3.40	3.50	3.60	3.70	3.80	3.90	4.00	4.10	4.20	4.30	4.40	4.50	4.60	4.70	4.80	4.90	5.00	5.10	x
0	.0334	.0302	.0273	.0247	.0224	.0202	.0183	.0166	.0150	.0136	.0123	.0111	.0101	.0091	.0082	.0074	.0067	.0061	0
1	.1135	.1057	.0984	.0915	.0850	.0789	.0733	.0679	.0630	.0583	.0540	.0500	.0462	.0427	.0395	.0365	.0337	.0311	1
2	.1929	.1850	.1771	.1692	.1615	.1539	.1465	.1393	.1323	.1254	.1188	.1125	.1063	.1005	.0948	.0894	.0842	.0793	2
3	.2186	.2158	.2125	.2087	.2046	.2001	.1954	.1904	.1852	.1798	.1743	.1687	.1631	.1574	.1517	.1460	.1404	.1348	3
4	.1858	.1888	.1912	.1931	.1944	.1951	.1954	.1951	.1944	.1933	.1917	.1898	.1875	.1849	.1820	.1789	.1755	.1719	4
5	.1264	.1322	.1377	.1429	.1477	.1522	.1563	.1600	.1633	.1662	.1687	.1708	.1725	.1738	.1747	.1753	.1755	.1753	5
6	.0716	.0771	.0826	.0881	.0936	.0989	.1042	.1093	.1143	.1191	.1237	.1281	.1323	.1362	.1398	.1432	.1462	.1490	6
7	.0348	.0385	.0425	.0466	.0508	.0551	.0595	.0640	.0686	.0732	.0778	.0824	.0869	.0914	.0959	.1002	.1044	.1086	7
8	.0148	.0169	.0191	.0215	.0241	.0269	.0298	.0328	.0360	.0393	.0428	.0463	.0500	.0537	.0575	.0614	.0653	.0692	8
9	.0056	.0066	.0076	.0089	.0102	.0116	.0132	.0150	.0168	.0188	.0209	.0232	.0255	.0281	.0307	.0334	.0363	.0392	9
10	.0019	.0023	.0028	.0033	.0039	.0045	.0053	.0061	.0071	.0081	.0092	.0104	.0118	.0132	.0147	.0164	.0181	.0200	10
11	.0006	.0007	.0009	.0011	.0013	.0016	.0019	.0023	.0027	.0032	.0037	.0043	.0049	.0056	.0064	.0073	.0082	.0093	11
12	.0002	.0002	.0003	.0003	.0004	.0005	.0006	.0008	.0009	.0011	.0013	.0016	.0019	.0022	.0026	.0030	.0034	.0039	12
13	.0000	.0001	.0001	.0001	.0001	.0002	.0002	.0002	.0003	.0004	.0005	.0006	.0007	.0008	.0009	.0011	.0013	.0015	13
14	.0000	.0000	.0000	.0000	.0000	.0000	.0001	.0001	.0001	.0001	.0001	.0002	.0002	.0003	.0003	.0004	.0005	.0006	14
15	.0000	.0000	.0000	.0000	.0000	.0000	.0000	.0000	.0000	.0000	.0000	.0001	.0001	.0001	.0001	.0001	.0002	.0002	15
16	.0000	.0000	.0000	.0000	.0000	.0000	.0000	.0000	.0000	.0000	.0000	.0000	.0000	.0000	.0000	.0000	.0000	.0001	16
17	.0000	.0000	.0000	.0000	.0000	.0000	.0000	.0000	.0000	.0000	.0000	.0000	.0000	.0000	.0000	.0000	.0000	.0000	17

Prob $(X = x)$

The main uses of the Poisson distribution are as an approximation to binomial distributions having large n and small p (for notation see page 4) and as a description of the occurrence of random events over time (or other continua). Individual probabilities are given on pages 14–16 for a wide range of values of the mean μ, and cumulative probabilities are obtained from the Poisson probability chart on page 17.

EXAMPLES: A production process is supposed to have a 1% rate of defectives. In a random sample of size eighty, what is the probability of there being (a) exactly two defectives, and (b) at least two defectives? The number X of defectives has a binomial distribution with $n = 80$ and $p = 0.01$; its mean μ is $np = 80 \times 0.01 = 0.8$. This distribution is well approximated by the Poisson distribution having the same mean, $\mu = 0.8$. So immediately we find (a) Prob $(X = 2) = 0.1438$. For (b) Prob $(X \geqslant 2)$ we can use the chart on page 17 directly. However this probability can also be found by noting that Prob $(X \geqslant 2) = 1 - \text{Prob}\,(X \leqslant 1) = 1 - \{\text{Prob}\,(X = 0) + \text{Prob}\,(X = 1)\} = 1 - \{0.4993 + 0.3595\} = 0.1912$, using the above table.

A binomial distribution with large n and a p-value close to 1 may also be dealt with by means of a Poisson approximation if the problem is re-expressed in terms of a small p-value. For example if a treatment has a 90% ($p = 0.9$) success-rate, what is the probability that exactly 95 out of 100 treated patients recover? This is the same as asking what is the probability that exactly 5 patients out of 100 fail to recover when the failure-rate is 10% or 0.1. That is we want Prob $(X = 5)$ in the binomial distribution with $n = 100$ and $p = 0.1$ which can be approximated by the Poisson distribution with mean $\mu = np = 100 \times 0.1 = 10.0$. From page 15, this probability is found to be 0.0378.

The Poisson distribution: individual probabilities

Prob $(X = x)$

x \ μ	5.20	5.30	5.40	5.50	5.60	5.70	5.80	5.90	6.00	6.10	6.20	6.30	6.40	6.50	6.60	6.70	6.80	6.90	μ / x
0	.0055	.0050	.0045	.0041	.0037	.0033	.0030	.0027	.0025	.0022	.0020	.0018	.0017	.0015	.0014	.0012	.0011	.0010	0
1	.0287	.0265	.0244	.0225	.0207	.0191	.0176	.0162	.0149	.0137	.0126	.0116	.0106	.0098	.0090	.0082	.0076	.0070	1
2	.0746	.0701	.0659	.0618	.0580	.0544	.0509	.0477	.0446	.0417	.0390	.0364	.0340	.0318	.0296	.0276	.0258	.0240	2
3	.1293	.1239	.1185	.1133	.1082	.1033	.0985	.0938	.0892	.0848	.0806	.0765	.0726	.0688	.0652	.0617	.0584	.0552	3
4	.1681	.1641	.1600	.1558	.1515	.1472	.1428	.1383	.1339	.1294	.1249	.1205	.1162	.1118	.1076	.1034	.0992	.0952	4
5	.1748	.1740	.1728	.1714	.1697	.1678	.1656	.1632	.1606	.1579	.1549	.1519	.1487	.1454	.1420	.1385	.1349	.1314	5
6	.1515	.1537	.1555	.1571	.1584	.1594	.1601	.1605	.1606	.1605	.1601	.1595	.1586	.1575	.1562	.1546	.1529	.1511	6
7	.1125	.1163	.1200	.1234	.1267	.1298	.1326	.1353	.1377	.1399	.1418	.1435	.1450	.1462	.1472	.1480	.1486	.1489	7
8	.0731	.0771	.0810	.0849	.0887	.0925	.0962	.0998	.1033	.1066	.1099	.1130	.1160	.1188	.1215	.1240	.1263	.1284	8
9	.0423	.0454	.0486	.0519	.0552	.0586	.0620	.0654	.0688	.0723	.0757	.0791	.0825	.0858	.0891	.0923	.0954	.0985	9
10	.0220	.0241	.0262	.0285	.0309	.0334	.0359	.0386	.0413	.0441	.0469	.0498	.0528	.0558	.0588	.0618	.0649	.0679	10
11	.0104	.0116	.0129	.0143	.0157	.0173	.0190	.0207	.0225	.0244	.0265	.0285	.0307	.0330	.0353	.0377	.0401	.0426	11
12	.0045	.0051	.0058	.0065	.0073	.0082	.0092	.0102	.0113	.0124	.0137	.0150	.0164	.0179	.0194	.0210	.0227	.0245	12
13	.0018	.0021	.0024	.0028	.0032	.0036	.0041	.0046	.0052	.0058	.0065	.0073	.0081	.0089	.0099	.0108	.0119	.0130	13
14	.0007	.0008	.0009	.0011	.0013	.0015	.0017	.0019	.0022	.0025	.0029	.0033	.0037	.0041	.0046	.0052	.0058	.0064	14
15	.0002	.0003	.0003	.0004	.0005	.0006	.0007	.0008	.0009	.0010	.0012	.0014	.0016	.0018	.0020	.0023	.0026	.0029	15
16	.0001	.0001	.0001	.0001	.0002	.0002	.0002	.0003	.0003	.0004	.0005	.0005	.0006	.0007	.0008	.0010	.0011	.0013	16
17	.0000	.0000	.0000	.0000	.0001	.0001	.0001	.0001	.0001	.0001	.0002	.0002	.0002	.0003	.0003	.0004	.0004	.0005	17
18	.0000	.0000	.0000	.0000	.0000	.0000	.0000	.0000	.0000	.0000	.0001	.0001	.0001	.0001	.0001	.0001	.0002	.0002	18
19	.0000	.0000	.0000	.0000	.0000	.0000	.0000	.0000	.0000	.0000	.0000	.0000	.0000	.0000	.0000	.0001	.0001	.0001	19
20	.0000	.0000	.0000	.0000	.0000	.0000	.0000	.0000	.0000	.0000	.0000	.0000	.0000	.0000	.0000	.0000	.0000	.0000	20

x \ μ	7.00	7.10	7.20	7.30	7.40	7.50	7.60	7.70	7.80	7.90	8.00	8.10	8.20	8.30	8.40	8.50	8.60	8.70	μ / x
0	.0009	.0008	.0007	.0007	.0006	.0006	.0005	.0005	.0004	.0004	.0003	.0003	.0003	.0002	.0002	.0002	.0002	.0002	0
1	.0064	.0059	.0054	.0049	.0045	.0041	.0038	.0035	.0032	.0029	.0027	.0025	.0023	.0021	.0019	.0017	.0016	.0014	1
2	.0223	.0208	.0194	.0180	.0167	.0156	.0145	.0134	.0125	.0116	.0107	.0100	.0092	.0086	.0079	.0074	.0068	.0063	2
3	.0521	.0492	.0464	.0438	.0413	.0389	.0366	.0345	.0324	.0305	.0286	.0269	.0252	.0237	.0222	.0208	.0195	.0183	3
4	.0912	.0874	.0836	.0799	.0764	.0729	.0696	.0663	.0632	.0602	.0573	.0544	.0517	.0491	.0466	.0443	.0420	.0398	4
5	.1277	.1241	.1204	.1167	.1130	.1094	.1057	.1021	.0986	.0951	.0916	.0882	.0840	.0816	.0784	.0752	.0722	.0692	5
6	.1490	.1468	.1445	.1420	.1394	.1367	.1339	.1311	.1282	.1252	.1221	.1191	.1160	.1128	.1097	.1066	.1034	.1003	6
7	.1490	.1489	.1486	.1481	.1474	.1465	.1454	.1442	.1428	.1413	.1396	.1378	.1358	.1338	.1317	.1294	.1271	.1247	7
8	.1304	.1321	.1337	.1351	.1363	.1373	.1381	.1388	.1392	.1395	.1396	.1395	.1392	.1388	.1382	.1375	.1366	.1356	8
9	.1014	.1042	.1070	.1096	.1121	.1144	.1167	.1187	.1207	.1224	.1241	.1256	.1269	.1280	.1290	.1299	.1306	.1311	9
10	.0710	.0740	.0770	.0800	.0829	.0858	.0887	.0914	.0941	.0967	.0993	.1017	.1040	.1063	.1084	.1104	.1123	.1140	10
11	.0452	.0478	.0504	.0531	.0558	.0585	.0613	.0640	.0667	.0695	.0722	.0749	.0776	.0802	.0828	.0853	.0878	.0902	11
12	.0263	.0283	.0303	.0323	.0344	.0366	.0388	.0411	.0434	.0457	.0481	.0505	.0530	.0555	.0579	.0604	.0629	.0654	12
13	.0142	.0154	.0168	.0181	.0196	.0211	.0227	.0243	.0260	.0278	.0296	.0315	.0334	.0354	.0374	.0395	.0416	.0438	13
14	.0071	.0078	.0086	.0095	.0104	.0113	.0123	.0134	.0145	.0157	.0169	.0182	.0196	.0210	.0225	.0240	.0256	.0272	14
15	.0033	.0037	.0041	.0046	.0051	.0057	.0062	.0069	.0075	.0083	.0090	.0098	.0107	.0116	.0126	.0136	.0147	.0158	15
16	.0014	.0016	.0019	.0021	.0024	.0026	.0030	.0033	.0037	.0041	.0045	.0050	.0055	.0060	.0066	.0072	.0079	.0086	16
17	.0006	.0007	.0008	.0009	.0010	.0012	.0013	.0015	.0017	.0019	.0021	.0024	.0026	.0029	.0033	.0036	.0040	.0044	17
18	.0002	.0003	.0003	.0004	.0004	.0005	.0006	.0006	.0007	.0008	.0009	.0011	.0012	.0014	.0015	.0017	.0019	.0021	18
19	.0001	.0001	.0001	.0001	.0002	.0002	.0002	.0003	.0003	.0003	.0004	.0005	.0005	.0006	.0007	.0008	.0009	.0010	19
20	.0000	.0000	.0000	.0001	.0001	.0001	.0001	.0001	.0001	.0001	.0002	.0002	.0002	.0002	.0003	.0003	.0004	.0004	20
21	.0000	.0000	.0000	.0000	.0000	.0000	.0000	.0000	.0000	.0001	.0001	.0001	.0001	.0001	.0001	.0001	.0002	.0002	21
22	.0000	.0000	.0000	.0000	.0000	.0000	.0000	.0000	.0000	.0000	.0000	.0000	.0000	.0000	.0000	.0001	.0001	.0001	22
23	.0000	.0000	.0000	.0000	.0000	.0000	.0000	.0000	.0000	.0000	.0000	.0000	.0000	.0000	.0000	.0000	.0000	.0000	23

x \ μ	8.80	8.90	9.00	9.10	9.20	9.30	9.40	9.50	9.60	9.70	9.80	9.90	10.00	10.50	11.00	11.50	12.00	12.50	μ / x
0	.0002	.0001	.0001	.0001	.0001	.0001	.0001	.0001	.0001	.0001	.0001	.0001	.0000	.0000	.0000	.0000	.0000	.0000	0
1	.0013	.0012	.0011	.0010	.0009	.0009	.0008	.0007	.0007	.0006	.0005	.0005	.0005	.0003	.0002	.0001	.0001	.0000	1
2	.0058	.0054	.0050	.0046	.0043	.0040	.0037	.0034	.0031	.0029	.0027	.0025	.0023	.0015	.0010	.0007	.0004	.0003	2
3	.0171	.0160	.0150	.0140	.0131	.0123	.0115	.0107	.0100	.0093	.0087	.0081	.0076	.0053	.0037	.0026	.0018	.0012	3
4	.0377	.0357	.0337	.0319	.0302	.0285	.0269	.0254	.0240	.0226	.0213	.0201	.0189	.0139	.0102	.0074	.0053	.0038	4
5	.0663	.0635	.0607	.0581	.0555	.0530	.0506	.0483	.0460	.0439	.0418	.0398	.0378	.0293	.0224	.0170	.0127	.0095	5
6	.0972	.0941	.0911	.0881	.0851	.0822	.0793	.0764	.0736	.0709	.0682	.0656	.0631	.0513	.0411	.0325	.0255	.0197	6
7	.1222	.1197	.1171	.1145	.1118	.1091	.1064	.1037	.1010	.0982	.0955	.0928	.0901	.0769	.0646	.0535	.0437	.0353	7
8	.1344	.1332	.1318	.1302	.1286	.1269	.1251	.1232	.1212	.1191	.1170	.1148	.1126	.1009	.0888	.0769	.0655	.0551	8
9	.1315	.1317	.1318	.1317	.1315	.1311	.1306	.1300	.1293	.1284	.1274	.1263	.1251	.1177	.1085	.0982	.0874	.0765	9
10	.1157	.1172	.1186	.1198	.1210	.1219	.1228	.1235	.1241	.1245	.1249	.1250	.1251	.1236	.1194	.1129	.1048	.0956	10
11	.0925	.0948	.0970	.0991	.1012	.1031	.1049	.1067	.1083	.1098	.1112	.1125	.1137	.1180	.1194	.1181	.1144	.1087	11
12	.0679	.0703	.0728	.0752	.0776	.0799	.0822	.0844	.0866	.0888	.0908	.0928	.0948	.1032	.1094	.1131	.1144	.1132	12
13	.0459	.0481	.0504	.0526	.0549	.0572	.0594	.0617	.0640	.0662	.0685	.0707	.0729	.0834	.0926	.1001	.1056	.1089	13
14	.0289	.0306	.0324	.0342	.0361	.0380	.0399	.0419	.0439	.0459	.0479	.0500	.0521	.0625	.0728	.0822	.0905	.0972	14
15	.0169	.0182	.0194	.0208	.0221	.0235	.0250	.0265	.0281	.0297	.0313	.0330	.0347	.0438	.0534	.0630	.0724	.0810	15
16	.0093	.0101	.0109	.0118	.0127	.0137	.0147	.0157	.0168	.0180	.0192	.0204	.0217	.0287	.0367	.0453	.0543	.0633	16
17	.0048	.0053	.0058	.0063	.0069	.0075	.0081	.0088	.0095	.0103	.0111	.0119	.0128	.0177	.0237	.0306	.0383	.0465	17
18	.0024	.0026	.0029	.0032	.0035	.0039	.0042	.0046	.0051	.0055	.0060	.0065	.0071	.0104	.0145	.0196	.0255	.0323	18
19	.0011	.0012	.0014	.0015	.0017	.0019	.0021	.0023	.0026	.0028	.0031	.0034	.0037	.0057	.0084	.0119	.0161	.0213	19
20	.0005	.0005	.0006	.0007	.0008	.0009	.0010	.0011	.0012	.0014	.0015	.0017	.0019	.0030	.0046	.0068	.0097	.0133	20
21	.0002	.0002	.0003	.0003	.0003	.0004	.0004	.0005	.0006	.0006	.0007	.0008	.0009	.0015	.0024	.0037	.0055	.0079	21
22	.0001	.0001	.0001	.0001	.0001	.0002	.0002	.0002	.0002	.0003	.0003	.0004	.0004	.0007	.0012	.0020	.0030	.0045	22
23	.0000	.0000	.0000	.0000	.0001	.0001	.0001	.0001	.0001	.0001	.0001	.0002	.0002	.0003	.0006	.0010	.0016	.0024	23
24	.0000	.0000	.0000	.0000	.0000	.0000	.0000	.0000	.0000	.0000	.0001	.0001	.0001	.0001	.0003	.0005	.0008	.0013	24
25	.0000	.0000	.0000	.0000	.0000	.0000	.0000	.0000	.0000	.0000	.0000	.0000	.0000	.0001	.0001	.0002	.0004	.0006	25
26	.0000	.0000	.0000	.0000	.0000	.0000	.0000	.0000	.0000	.0000	.0000	.0000	.0000	.0000	.0000	.0001	.0002	.0003	26
27	.0000	.0000	.0000	.0000	.0000	.0000	.0000	.0000	.0000	.0000	.0000	.0000	.0000	.0000	.0000	.0000	.0001	.0001	27
28	.0000	.0000	.0000	.0000	.0000	.0000	.0000	.0000	.0000	.0000	.0000	.0000	.0000	.0000	.0000	.0000	.0000	.0001	28
29	.0000	.0000	.0000	.0000	.0000	.0000	.0000	.0000	.0000	.0000	.0000	.0000	.0000	.0000	.0000	.0000	.0000	.0000	29

Prob $(X = x)$

Prob $(X = x)$

μ / x	13.00	13.50	14.00	14.50	15.00	16.00	17.00	18.00	19.00	20.00	21.00	22.00	23.00	24.00	25.00	30.00	40.00	50.00	x
0	.0000	.0000	.0000	.0000	.0000	.0000	.0000	.0000	.0000	.0000	.0000	.0000	.0000	.0000	.0000	.0000	.0000	.0000	0
1	.0000	.0000	.0000	.0000	.0000	.0000	.0000	.0000	.0000	.0000	.0000	.0000	.0000	.0000	.0000	.0000	.0000	.0000	1
2	.0002	.0001	.0001	.0001	.0000	.0000	.0000	.0000	.0000	.0000	.0000	.0000	.0000	.0000	.0000	.0000	.0000	.0000	2
3	.0008	.0006	.0004	.0003	.0002	.0001	.0000	.0000	.0000	.0000	.0000	.0000	.0000	.0000	.0000	.0000	.0000	.0000	3
4	.0027	.0019	.0013	.0009	.0006	.0003	.0001	.0001	.0000	.0000	.0000	.0000	.0000	.0000	.0000	.0000	.0000	.0000	4
5	.0070	.0051	.0037	.0027	.0019	.0010	.0005	.0002	.0001	.0001	.0000	.0000	.0000	.0000	.0000	.0000	.0000	.0000	5
6	.0152	.0115	.0087	.0065	.0048	.0026	.0014	.0007	.0004	.0002	.0001	.0000	.0000	.0000	.0000	.0000	.0000	.0000	6
7	.0281	.0222	.0174	.0135	.0104	.0060	.0034	.0019	.0010	.0005	.0003	.0001	.0001	.0000	.0000	.0000	.0000	.0000	7
8	.0457	.0375	.0304	.0244	.0194	.0120	.0072	.0042	.0024	.0013	.0007	.0004	.0002	.0001	.0001	.0000	.0000	.0000	8
9	.0661	.0563	.0473	.0394	.0324	.0213	.0135	.0083	.0050	.0029	.0017	.0009	.0005	.0003	.0001	.0000	.0000	.0000	9
10	.0859	.0760	.0663	.0571	.0486	.0341	.0230	.0150	.0095	.0058	.0035	.0020	.0012	.0007	.0004	.0000	.0000	.0000	10
11	.1015	.0932	.0844	.0753	.0663	.0496	.0355	.0245	.0164	.0106	.0067	.0041	.0024	.0014	.0008	.0000	.0000	.0000	11
12	.1099	.1049	.0984	.0910	.0829	.0661	.0504	.0368	.0259	.0176	.0116	.0075	.0047	.0029	.0017	.0001	.0000	.0000	12
13	.1099	.1089	.1060	.1014	.0956	.0814	.0658	.0509	.0378	.0271	.0188	.0127	.0083	.0053	.0033	.0002	.0000	.0000	13
14	.1021	.1050	.1060	.1051	.1024	.0930	.0800	.0655	.0514	.0387	.0282	.0199	.0136	.0091	.0059	.0005	.0000	.0000	14
15	.0885	.0945	.0989	.1016	.1024	.0992	.0906	.0786	.0650	.0516	.0395	.0292	.0209	.0146	.0099	.0010	.0000	.0000	15
16	.0719	.0798	.0866	.0920	.0960	.0992	.0963	.0884	.0772	.0646	.0518	.0401	.0301	.0219	.0155	.0019	.0000	.0000	16
17	.0550	.0633	.0713	.0785	.0847	.0934	.0963	.0936	.0863	.0760	.0640	.0520	.0407	.0309	.0227	.0034	.0000	.0000	17
18	.0397	.0475	.0554	.0632	.0706	.0830	.0909	.0936	.0911	.0844	.0747	.0635	.0520	.0412	.0316	.0057	.0000	.0000	18
19	.0272	.0337	.0409	.0483	.0557	.0699	.0814	.0887	.0911	.0888	.0826	.0735	.0629	.0520	.0415	.0089	.0001	.0000	19
20	.0177	.0228	.0286	.0350	.0418	.0559	.0692	.0798	.0866	.0888	.0867	.0809	.0724	.0624	.0519	.0134	.0002	.0000	20
21	.0109	.0146	.0191	.0242	.0299	.0426	.0560	.0684	.0783	.0846	.0867	.0847	.0793	.0713	.0618	.0192	.0004	.0000	21
22	.0065	.0090	.0121	.0159	.0204	.0310	.0433	.0560	.0676	.0769	.0828	.0847	.0829	.0778	.0702	.0261	.0007	.0000	22
23	.0037	.0053	.0074	.0100	.0133	.0216	.0320	.0438	.0559	.0669	.0756	.0810	.0829	.0812	.0763	.0341	.0012	.0000	23
24	.0020	.0030	.0043	.0061	.0083	.0144	.0226	.0328	.0442	.0557	.0661	.0743	.0794	.0812	.0795	.0426	.0019	.0000	24
25	.0010	.0016	.0024	.0035	.0050	.0092	.0154	.0237	.0336	.0446	.0555	.0654	.0731	.0779	.0795	.0511	.0031	.0000	25
26	.0005	.0008	.0013	.0020	.0029	.0057	.0101	.0164	.0246	.0343	.0449	.0553	.0646	.0719	.0765	.0590	.0047	.0001	26
27	.0002	.0004	.0007	.0011	.0016	.0034	.0063	.0109	.0173	.0254	.0349	.0451	.0551	.0639	.0708	.0655	.0070	.0001	27
28	.0001	.0002	.0003	.0005	.0009	.0019	.0038	.0070	.0117	.0181	.0262	.0354	.0452	.0548	.0632	.0702	.0100	.0002	28
29	.0001	.0001	.0002	.0003	.0004	.0011	.0023	.0044	.0077	.0125	.0190	.0269	.0359	.0453	.0545	.0726	.0138	.0004	29
30	.0000	.0000	.0001	.0001	.0002	.0006	.0013	.0026	.0049	.0083	.0133	.0197	.0275	.0363	.0454	.0726	.0185	.0007	30
31	.0000	.0000	.0000	.0001	.0001	.0003	.0007	.0015	.0030	.0054	.0090	.0140	.0204	.0281	.0366	.0703	.0238	.0011	31
32	.0000	.0000	.0000	.0000	.0001	.0001	.0004	.0009	.0018	.0034	.0059	.0096	.0147	.0211	.0286	.0659	.0298	.0017	32
33	.0000	.0000	.0000	.0000	.0000	.0001	.0002	.0005	.0010	.0020	.0038	.0064	.0102	.0153	.0217	.0599	.0361	.0026	33
34	.0000	.0000	.0000	.0000	.0000	.0000	.0001	.0002	.0006	.0012	.0023	.0041	.0069	.0108	.0159	.0529	.0425	.0038	34
35	.0000	.0000	.0000	.0000	.0000	.0000	.0000	.0001	.0003	.0007	.0014	.0026	.0045	.0074	.0114	.0453	.0485	.0054	35
36	.0000	.0000	.0000	.0000	.0000	.0000	.0000	.0001	.0002	.0004	.0008	.0016	.0029	.0049	.0079	.0378	.0539	.0075	36
37	.0000	.0000	.0000	.0000	.0000	.0000	.0000	.0000	.0001	.0002	.0005	.0009	.0018	.0032	.0053	.0306	.0583	.0102	37
38	.0000	.0000	.0000	.0000	.0000	.0000	.0000	.0000	.0000	.0001	.0003	.0005	.0011	.0020	.0035	.0242	.0614	.0134	38
39	.0000	.0000	.0000	.0000	.0000	.0000	.0000	.0000	.0000	.0001	.0001	.0003	.0006	.0012	.0023	.0186	.0629	.0172	39
40	.0000	.0000	.0000	.0000	.0000	.0000	.0000	.0000	.0000	.0000	.0001	.0002	.0004	.0007	.0014	.0139	.0629	.0215	40
41	.0000	.0000	.0000	.0000	.0000	.0000	.0000	.0000	.0000	.0000	.0000	.0001	.0002	.0004	.0009	.0102	.0614	.0262	41
42	.0000	.0000	.0000	.0000	.0000	.0000	.0000	.0000	.0000	.0000	.0000	.0000	.0001	.0003	.0005	.0073	.0585	.0312	42
43	.0000	.0000	.0000	.0000	.0000	.0000	.0000	.0000	.0000	.0000	.0000	.0000	.0001	.0001	.0003	.0051	.0544	.0363	43
44	.0000	.0000	.0000	.0000	.0000	.0000	.0000	.0000	.0000	.0000	.0000	.0000	.0000	.0001	.0002	.0035	.0495	.0412	44
45	.0000	.0000	.0000	.0000	.0000	.0000	.0000	.0000	.0000	.0000	.0000	.0000	.0000	.0000	.0001	.0023	.0440	.0458	45
46	.0000	.0000	.0000	.0000	.0000	.0000	.0000	.0000	.0000	.0000	.0000	.0000	.0000	.0000	.0001	.0015	.0382	.0498	46
47	.0000	.0000	.0000	.0000	.0000	.0000	.0000	.0000	.0000	.0000	.0000	.0000	.0000	.0000	.0000	.0010	.0325	.0530	47
48	.0000	.0000	.0000	.0000	.0000	.0000	.0000	.0000	.0000	.0000	.0000	.0000	.0000	.0000	.0000	.0006	.0271	.0552	48
49	.0000	.0000	.0000	.0000	.0000	.0000	.0000	.0000	.0000	.0000	.0000	.0000	.0000	.0000	.0000	.0004	.0221	.0563	49
50	.0000	.0000	.0000	.0000	.0000	.0000	.0000	.0000	.0000	.0000	.0000	.0000	.0000	.0000	.0000	.0002	.0177	.0563	50

	25.00	30.00	40.00	50.00	x
	.0000	.0001	.0139	.0552	51
	.0000	.0001	.0107	.0531	52
	.0000	.0000	.0081	.0501	53
	.0000	.0000	.0060	.0464	54
	.0000	.0000	.0043	.0422	55
	.0000	.0000	.0031	.0376	56
	.0000	.0000	.0022	.0330	57
	.0000	.0000	.0015	.0285	58
	.0000	.0000	.0010	.0241	59
	.0000	.0000	.0007	.0201	60
	.0000	.0000	.0004	.0165	61
	.0000	.0000	.0003	.0133	62
	.0000	.0000	.0002	.0105	63
	.0000	.0000	.0001	.0082	64
	.0000	.0000	.0001	.0063	65
	.0000	.0000	.0000	.0048	66
	.0000	.0000	.0000	.0036	67
	.0000	.0000	.0000	.0026	68
	.0000	.0000	.0000	.0019	69
	.0000	.0000	.0000	.0014	70
	.0000	.0000	.0000	.0010	71
	.0000	.0000	.0000	.0007	72
	.0000	.0000	.0000	.0005	73
	.0000	.0000	.0000	.0003	74
	.0000	.0000	.0000	.0002	75
	.0000	.0000	.0000	.0001	76
	.0000	.0000	.0000	.0001	77
	.0000	.0000	.0000	.0001	78
	.0000	.0000	.0000	.0000	79

The Poisson probability chart on page 17 gives cumulative probabilities of the form Prob $(X \geqslant x)$ where X has a Poisson distribution with mean μ in the range $0.01 \leqslant \mu \leqslant 100$. To find such a probability, locate the appropriate value of μ on the right-hand vertical axis, trace back along the horizontal to the line or curve labelled with the desired value of x, and read off the probability on the horizontal axis. The horizontal scale is designed to give most accuracy in the tails of the distribution, i.e. where the probabilities are close to 0 or 1, and the vertical scale has been devised to make the curves almost linear.

EXAMPLES: A production process is supposed to have a 1% rate of defectives. In a random sample of size eighty, what is the probability of there being at least two defectives? This question has already been answered on p. 14 using individual probabilities. Here we may read off the probability directly, following the above directions with $\mu = 0.8$ and $x = 2$, giving Prob $(X \geqslant 2) = 0.19$. Obviously, accuracy may be somewhat limited when using the chart.

Probabilities of events such as $X \leqslant 2$ can also be easily found. For Prob $(X \leqslant 2) = 1 -$ Prob $(X \geqslant 3)$, and Prob $(X \geqslant 3)$ is seen to be just less than 0.05, say 0.048, giving Prob $(X \leqslant 2) = 1 - 0.048 = 0.952$.

As a final example, suppose the number X of serious road accidents per week in a certain region has a Poisson distribution with mean $\mu = 2.0$. What is the probability of there being no more than three accidents in a particular week? This again can be calculated using either individual probabilities or the chart. From page 14, the probabilities of 0, 1, 2 or 3 accidents are respectively 0.1353, 0.2707, 0.2707 and 0.1804, and adding these we have Prob $(X \leqslant 3) = 0.8571$. Using the chart, since Prob $(X \leqslant 3) = 1 -$ Prob $(X \geqslant 4)$, we obtain Prob $(X \leqslant 3) = 1 - 0.14 = 0.86$.

Poisson probability chart (cumulative probabilities)

$$\text{Prob}\,(X \geqslant x) = \sum_{r=x}^{\infty} e^{-\mu} \cdot \frac{\mu^r}{r!}$$

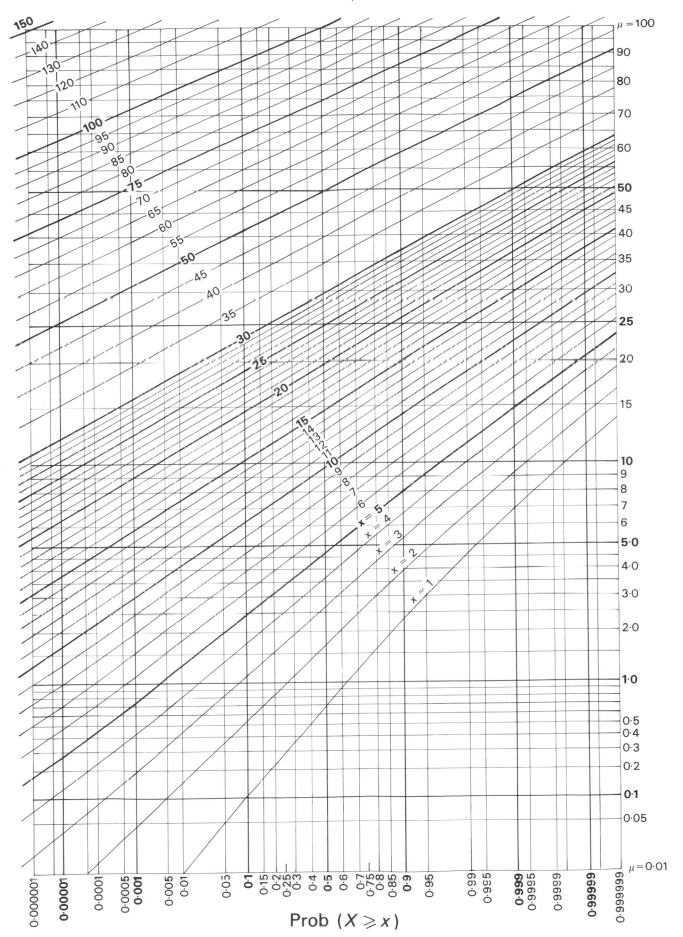

Prob ($X \geqslant x$)

For description, see page 16.

Probabilities and ordinates in the normal distribution

$$\phi(z) = \frac{1}{\sqrt{2\pi}} e^{-\frac{1}{2}z^2}; \quad \Phi(z) = \text{Prob}\,(Z \leqslant z) = \int_{-\infty}^{z} \phi(t)\,\mathrm{d}t$$

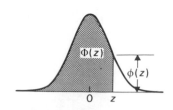

$\phi(z)$	z	0	1	2	3	4	5	6	7	8	9	1	2	3	4	5	6	7	8	9
0.0^8608	−6.0	0.0^9987	0^9928	0^9872	0^9820	0^9771	0^9724	0^9681	0^9640	0^9601	0^9565									
0.0^7110	−5.9	0.0^8182	0^8171	0^8161	0^8151	0^8143	0^8134	0^8126	0^8119	0^8112	0^8105									
0.0^7198	−5.8	0.0^8332	0^8312	0^8294	0^8277	0^8261	0^8246	0^8231	0^8218	0^8205	0^8193									
0.0^7351	−5.7	0.0^8599	0^8565	0^8533	0^8502	0^8473	0^8446	0^8421	0^8396	0^8374	0^8352									
0.0^7618	−5.6	0.0^7107	0^7101	0^8955	0^8901	0^8850	0^8802	0^8757	0^8714	0^8673	0^8635									
0.0^6108	−5.5	0.0^7190	0^7179	0^7169	0^7160	0^7151	0^7143	0^7135	0^7127	0^7120	0^7114									
0.0^6186	−5.4	0.0^7333	0^7315	0^7298	0^7282	0^7266	0^7252	0^7238	0^7225	0^7213	0^7201									
0.0^6317	−5.3	0.0^7579	0^7548	0^7519	0^7491	0^7465	0^7440	0^7416	0^7394	0^7372	0^7352									
0.0^6536	−5.2	0.0^7996	0^7944	0^7895	0^7848	0^7803	0^7760	0^7720	0^7682	0^7646	0^7612									
0.0^6897	−5.1	0.0^6170	0^6161	0^6153	0^6145	0^6137	0^6130	0^6123	0^6117	0^6111	0^6105									
0.0^5149	−5.0	0.0^6287	0^6272	0^6258	0^6245	0^6233	0^6221	0^6210	0^6199	0^6189	0^6179									
0.0^5244	−4.9	0.0^6479	0^6455	0^6433	0^6411	0^6391	0^6371	0^6352	0^6335	0^6318	0^6302									
0.0^5396	−4.8	0.0^6793	0^6755	0^6718	0^6683	0^6649	0^6617	0^6587	0^6558	0^6530	0^6504									
0.0^5637	−4.7	0.0^5130	0^5124	0^5118	0^5112	0^5107	0^5102	0^6968	0^6921	0^6876	0^6834									
0.0^4101	−4.6	0.0^5211	0^5201	0^5192	0^5183	0^5174	0^5166	0^5158	0^5151	0^5143	0^5137									
0.0^4160	−4.5	0.0^5340	0^5324	0^5309	0^5295	0^5281	0^5268	0^5256	0^5244	0^5232	0^5222									
0.0^4249	−4.4	0.0^5541	0^5517	0^5494	0^5471	0^5450	0^5429	0^5410	0^5391	0^5373	0^5356									
0.0^4385	−4.3	0.0^5854	0^5816	0^5780	0^5746	0^5712	0^5681	0^5650	0^5621	0^5593	0^5567									
0.0^4589	−4.2	0.0^4133	0^4128	0^4122	0^4117	0^4112	0^4107	0^4102	0^5977	0^5934	0^5893									
0.0^4893	−4.1	0.0^4207	0^4198	0^4189	0^4181	0^4174	0^4166	0^4159	0^4152	0^4146	0^4139									
0.0^3134	−4.0	0.0^4317	0^4304	0^4291	0^4279	0^4267	0^4256	0^4245	0^4235	0^4225	0^4216									
0.0^3199	−3.9	0.0^4481	0^4461	0^4443	0^4425	0^4407	0^4391	0^4375	0^4359	0^4345	0^4330									
0.0^3292	−3.8	0.0^4723	0^4695	0^4667	0^4641	0^4615	0^4591	0^4567	0^4544	0^4522	0^4501									
0.0^3425	−3.7	0.0^3108	0^3104	0^4996	0^4957	0^4920	0^4884	0^4850	0^4816	0^4784	0^4753									
0.0^3612	−3.6	0.0^3159	0^3153	0^3147	0^3142	0^3136	0^3131	0^3126	0^3121	0^3117	0^3112									
0.0^3873	−3.5	0.0^3233	0^3224	0^3216	0^3208	0^3200	0^3193	0^3185	0^3178	0^3172	0^3165									
0.00123	−3.4	0.0^3337	0^3325	0^3313	0^3302	0^3291	0^3280	0^3270	0^3260	0^3251	0^3242									
0.00172	−3.3	0.0^3483	0^3466	0^3450	0^3434	0^3419	0^3404	0^3390	0^3376	0^3362	0^3349									
0.00238	−3.2	0.0^3687	0^3664	0^3641	0^3619	0^3598	0^3577	0^3557	0^3538	0^3519	0^3501	1	2	3	4	5	6	7	8	9
0.00327	−3.1	0.0^3968	0^3935	0^3904	0^3874	0^3845	0^3816	0^3789	0^3762	0^3736	0^3711									
0.00443	−3.0	0.00135	00131	00126	00122	00118	00114	00111	00107	00104	00100	0	1	1	2	2	2	3	3	3
0.00595	−2.9	0.00187	00181	00175	00169	00164	00159	00154	00149	00144	00139	1	1	2	2	3	3	4	4	5
0.00792	−2.8	0.00256	00248	00240	00233	00226	00219	00212	00205	00199	00193	1	1	2	3	3	4	5	6	6
0.0104	−2.7	0.00347	00336	00326	00317	00307	00298	00289	00280	00272	00264	1	2	3	4	5	5	6	7	8
0.0136	−2.6	0.00466	00453	00440	00427	00415	00402	00391	00379	00368	00357	1	2	4	5	6	7	8	10	11
0.0175	−2.5	0.00621	00604	00587	00570	00554	00539	00523	00508	00494	00480	2	3	5	6	8	9	11	12	14
0.0224	−2.4	0.00820	00798	00776	00755	00734	00714	00695	00676	00657	00639	2	4	6	8	10	12	14	16	18
0.0283	−2.3	0.0107	0104	0102	0099	0096	0094	0091	0089	0087	0084	0	0	1	1	1	2	2	2	2
0.0355	−2.2	0.0139	0136	0132	0129	0125	0122	0119	0116	0113	0110	0	1	1	1	2	2	2	3	3
0.0440	−2.1	0.0179	0174	0170	0166	0162	0158	0154	0150	0146	0143	0	1	1	2	2	2	3	3	4
0.0540	−2.0	0.0228	0222	0217	0212	0207	0202	0197	0192	0188	0183	0	1	1	2	2	3	3	4	4
0.0656	−1.9	0.0287	0281	0274	0268	0262	0256	0250	0244	0239	0233	1	1	2	2	3	4	4	5	5
0.0790	−1.8	0.0359	0351	0344	0336	0329	0322	0314	0307	0301	0294	1	1	2	3	4	4	5	6	6
0.0940	−1.7	0.0446	0436	0427	0418	0409	0401	0392	0384	0375	0367	1	2	3	3	4	5	6	7	8
0.1109	−1.6	0.0548	0537	0526	0516	0505	0495	0485	0475	0465	0455	1	2	3	4	5	6	7	8	9
0.1295	−1.5	0.0668	0655	0643	0630	0618	0606	0594	0582	0571	0559	1	2	4	5	6	7	8	10	11
0.1497	−1.4	0.0808	0793	0778	0764	0749	0735	0721	0708	0694	0681	1	3	4	6	7	8	10	11	13
0.1714	−1.3	0.0968	0951	0934	0918	0901	0885	0869	0853	0838	0823	2	3	5	6	8	10	11	13	14
0.1942	−1.2	0.1151	1131	1112	1093	1075	1056	1038	1020	1003	0985	2	4	5	7	9	11	13	15	16
0.2179	−1.1	0.1357	1335	1314	1292	1271	1251	1230	1210	1190	1170	2	4	6	8	10	12	14	16	19
0.2420	−1.0	0.1587	1562	1539	1515	1492	1469	1446	1423	1401	1379	2	5	7	9	12	14	16	18	21
0.2661	−0.9	0.1841	1814	1788	1762	1736	1711	1685	1660	1635	1611	3	5	8	10	13	15	18	20	23
0.2897	−0.8	0.2119	2090	2061	2033	2005	1977	1949	1922	1894	1867	3	6	8	11	14	17	19	22	25
0.3123	−0.7	0.2420	2389	2358	2327	2296	2266	2236	2206	2177	2148	3	6	9	12	15	18	21	24	27
0.3332	−0.6	0.2743	2709	2676	2643	2611	2578	2546	2514	2483	2451	3	6	10	13	16	19	23	26	29
0.3521	−0.5	0.3085	3050	3015	2981	2946	2912	2877	2843	2810	2776	3	7	10	14	17	21	24	27	31
0.3683	−0.4	0.3446	3409	3372	3336	3300	3264	3228	3192	3156	3121	4	7	11	14	18	22	25	29	32
0.3814	−0.3	0.3821	3783	3745	3707	3669	3632	3594	3557	3520	3483	4	8	11	15	19	22	26	30	34
0.3910	−0.2	0.4207	4168	4129	4090	4052	4013	3974	3936	3897	3859	4	8	12	15	19	23	27	31	35
0.3970	−0.1	0.4602	4562	4522	4483	4443	4404	4364	4325	4286	4247	4	8	12	16	20	24	28	32	36
0.3989	−0.0	0.5000	4960	4920	4880	4840	4801	4761	4721	4681	4641	4	8	12	16	20	24	28	32	36
$\phi(z)$	z	0	1	2	3	4	5	6	7	8	9	1	2	3	4	5	6	7	8	9

SUBTRACT PROPORTIONAL PARTS (columns 1–9)

SUBTRACT PROPORTIONAL PARTS

The superscript in numbers such as 0.0^8182 indicates a number of zeros, thus: $0.0^8182 = 0.000\,000\,00182$, and $0.0^3483 = 0.000\,483$.

Proportional parts have not been given in this region because they would not be of sufficient accuracy.

The left-hand column gives the ordinate $\phi(z) = e^{-\frac{1}{2}z^2}/\sqrt{2\pi}$ of the standard normal distribution (i.e. the normal distribution having mean 0 and standard deviation 1), z being listed in the second column. The rest of the table gives $\Phi(z) = \int_{-\infty}^{z} \phi(t)\,\mathrm{d}t = \text{Prob}\,(Z \leqslant z)$, where Z is a random variable having the standard normal distribution. Locate z, expressed to its first decimal place in the second column, and its second decimal place along the top or bottom horizontal: the corresponding table entry is $\Phi(z)$. Proportional parts are given for the third decimal place of z in part of the table. These proportional parts should be subtracted if $z < 0$ and added if $z > 0$.

EXAMPLES: $\Phi(-1.2) = \text{Prob}\,(Z \leqslant -1.2) = 0.1151$; $\Phi(-1.23) = 0.1093$; $\Phi(-1.234) = 0.1086$.

Probabilities and ordinates in the normal distribution

												ADD PROPORTIONAL PARTS								
$\phi(z)$	z	0	1	2	3	4	5	6	7	8	9	1	2	3	4	5	6	7	8	9
0.3989	0.0	0.5000	5040	5080	5120	5160	5199	5239	5279	5319	5359	4	8	12	16	20	24	28	32	36
0.3970	0.1	0.5398	5438	5478	5517	5557	5596	5636	5675	5714	5753	4	8	12	16	20	24	28	32	36
0.3910	0.2	0.5793	5832	5871	5910	5948	5987	6026	6064	6103	6141	4	8	12	15	19	23	27	31	35
0.3814	0.3	0.6179	6217	6255	6293	6331	6368	6406	6443	6480	6517	4	8	11	15	19	22	26	30	34
0.3683	0.4	0.6554	6591	6628	6664	6700	6736	6772	6808	6844	6879	4	7	11	14	18	22	25	29	32
0.3521	0.5	0.6915	6950	6985	7019	7054	7088	7123	7157	7190	7224	3	7	10	14	17	21	24	27	31
0.3332	0.6	0.7257	7291	7324	7357	7389	7422	7454	7486	7517	7549	3	6	10	13	16	19	23	26	29
0.3123	0.7	0.7580	7611	7642	7673	7704	7734	7764	7794	7823	7852	3	6	9	12	15	18	21	24	27
0.2897	0.8	0.7881	7910	7939	7967	7995	8023	8051	8078	8106	8133	3	6	8	11	14	17	19	22	25
0.2661	0.9	0.8159	8186	8212	8238	8264	8289	8315	8340	8365	8389	3	5	8	10	13	15	18	20	23
0.2420	1.0	0.8413	8438	8461	8485	8508	8531	8554	8577	8599	8621	2	5	7	9	12	14	16	18	21
0.2179	1.1	0.8643	8665	8686	8708	8729	8749	8770	8790	8810	8830	2	4	6	8	10	12	14	16	19
0.1942	1.2	0.8849	8869	8888	8907	8925	8944	8962	8980	8997	9015	2	4	5	7	9	11	13	15	16
0.1714	1.3	0.9032	9049	9066	9082	9099	9115	9131	9147	9162	9177	2	3	5	6	8	10	11	13	14
0.1497	1.4	0.9192	9207	9222	9236	9251	9265	9279	9292	9306	9319	1	3	4	6	7	8	10	11	13
0.1295	1.5	0.9332	9345	9357	9370	9382	9394	9406	9418	9429	9441	1	2	4	5	6	7	8	10	11
0.1109	1.6	0.9452	9463	9474	9484	9495	9505	9515	9525	9535	9545	1	2	3	4	5	6	7	8	9
0.0940	1.7	0.9554	9564	9573	9582	9591	9599	9608	9616	9625	9633	1	2	3	3	4	5	6	7	8
0.0790	1.8	0.9641	9649	9656	9664	9671	9678	9686	9693	9699	9706	1	1	2	3	4	4	5	6	6
0.0656	1.9	0.9713	9719	9726	9732	9738	9744	9750	9756	9761	9767	1	1	2	2	3	4	4	5	5
0.0540	2.0	0.9772	9778	9783	9788	9793	9798	9803	9808	9812	9817	0	1	1	2	2	3	3	4	4
0.0440	2.1	0.9821	9826	9830	9834	9838	9842	9846	9850	9854	9857	0	1	1	2	2	2	3	3	4
0.0355	2.2	0.9861	9864	9868	9871	9875	9878	9881	9884	9887	9890	0	1	1	1	2	2	2	3	3
0.0283	2.3	0.9893	9896	9898	9901	9904	9906	9909	9911	9913	9916	0	0	1	1	1	2	2	2	2
0.0224	2.4	0.99180	99202	99224	99245	99266	99286	99305	99324	99343	99361	2	4	6	8	10	12	14	16	18
0.0175	2.5	0.99379	99396	99413	99430	99446	99461	99477	99492	99506	99520	2	3	5	6	8	9	11	12	14
0.0136	2.6	0.99534	99547	99560	99573	99585	99598	99609	99621	99632	99643	1	2	4	5	6	7	8	10	11
0.0104	2.7	0.99653	99664	99674	99683	99693	99702	99711	99720	99728	99736	1	2	3	4	5	5	6	7	8
0.00792	2.8	0.99744	99752	99760	99767	99774	99781	99788	99795	99801	99807	1	1	2	3	3	4	5	6	6
0.00595	2.9	0.99813	99819	99825	99831	99836	99841	99846	99851	99856	99861	1	1	2	2	3	3	4	4	5
0.00443	3.0	0.99865	99869	99874	99878	99882	99886	99889	99893	99896	99900	0	1	1	2	2	2	3	3	3

												1	2	3	4	5	6	7	8	9
0.00327	3.1	0.9^3032	9^3065	9^3096	9^3126	9^3155	9^3184	9^3211	9^3238	9^3264	9^3289									
0.00238	3.2	0.9^3313	9^3336	9^3359	9^3381	9^3402	9^3423	9^3443	9^3462	9^3481	9^3499									
0.00172	3.3	0.9^3517	9^3534	9^3550	9^3566	9^3581	9^3596	9^3610	9^3624	9^3638	9^3651									
0.00123	3.4	0.9^3663	9^3675	9^3687	9^3698	9^3709	9^3720	9^3730	9^3740	9^3749	9^3758									
0.0^3873	3.5	0.9^3767	9^3776	9^3784	9^3792	9^3800	9^3807	9^3815	9^3822	9^3828	9^3835									
0.0^3612	3.6	0.9^3841	9^3847	9^3853	9^3858	9^3864	9^3869	9^3874	9^3879	9^3883	9^3888									
0.0^3425	3.7	0.9^3892	9^3896	9^4004	9^4043	9^4080	9^4116	9^4150	9^4184	9^4216	9^4247									
0.0^3292	3.8	0.9^4277	9^4305	9^4333	9^4359	9^4385	9^4409	9^4433	9^4456	9^4478	9^4499									
0.0^3199	3.9	0.9^4519	9^4539	9^4557	9^4575	9^4593	9^4609	9^4625	9^4641	9^4655	9^4670									
0.0^3134	4.0	0.9^4683	9^4696	9^4709	9^4721	9^4733	9^4744	9^4755	9^4765	9^4775	9^4784									
0.0^4893	4.1	0.9^4793	9^4802	9^4811	9^4819	9^4826	9^4834	9^4841	9^4848	9^4854	9^4861									
0.0^4589	4.2	0.9^4867	9^4872	9^4878	9^4883	9^4888	9^4893	9^4898	9^5023	9^5066	9^5107									
0.0^4385	4.3	0.9^5146	9^5184	9^5220	9^5254	9^5288	9^5319	9^5350	9^5379	9^5407	9^5433									
0.0^4249	4.4	0.9^5459	9^5483	9^5506	9^5529	9^5550	9^5571	9^5590	9^5609	9^5627	9^5644									
0.0^4160	4.5	0.9^5660	9^5676	9^5691	9^5705	9^5719	9^5732	9^5744	9^5756	9^5768	9^5778									
0.0^4101	4.6	0.9^5789	9^5799	9^5808	9^5817	9^5826	9^5834	9^5842	9^5849	9^5857	9^5863									
0.0^5637	4.7	0.9^5870	9^5876	9^5882	9^5888	9^5893	9^5898	9^6032	9^6079	9^6124	9^6166									
0.0^5396	4.8	0.9^6207	9^6245	9^6282	9^6317	9^6351	9^6383	9^6413	9^6442	9^6470	9^6496									
0.0^5244	4.9	0.9^6521	9^6545	9^6567	9^6589	9^6609	9^6629	9^6648	9^6665	9^6682	9^6698									
0.0^5149	5.0	0.9^6713	9^6728	9^6742	9^6755	9^6767	9^6779	9^6790	9^6801	9^6811	9^6821									
0.0^6897	5.1	0.9^6830	9^6839	9^6847	9^6855	9^6863	9^6870	9^6877	9^6883	9^6889	9^6895									
0.0^6536	5.2	0.9^7004	9^7056	9^7105	9^7152	9^7197	9^7240	9^7280	9^7318	9^7354	9^7388									
0.0^6317	5.3	0.9^7421	9^7452	9^7481	9^7509	9^7535	9^7560	9^7584	9^7606	9^7628	9^7648									
0.0^6186	5.4	0.9^7667	9^7685	9^7702	9^7718	9^7734	9^7748	9^7762	9^7775	9^7787	9^7799									
0.0^6108	5.5	0.9^7010	9^7021	9^7031	9^7040	9^7040	9^7057	9^7866	9^7873	9^7880	9^7886									
0.0^7618	5.6	0.9^7893	9^7899	9^8045	9^8099	9^8150	9^8198	9^8243	9^8286	9^8327	9^8365									
0.0^7351	5.7	0.9^8401	9^8435	9^8467	9^8498	9^8527	9^8554	9^8579	9^8604	9^8626	9^8648									
0.0^7198	5.8	0.9^8668	9^8688	9^8706	9^8723	9^8739	9^8754	9^8769	9^8782	9^8795	9^8807									
0.0^7110	5.9	0.9^8818	9^8829	9^8839	9^8849	9^8857	9^8866	9^8874	9^8881	9^8888	9^8895									
0.0^8608	6.0	0.9^9013	9^9072	9^9128	9^9180	9^9229	9^9276	9^9319	9^9360	9^9399	9^9435									
$\phi(z)$	z	0	1	2	3	4	5	6	7	8	9									

ADD PROPORTIONAL PARTS

The superscript in numbers such as 0.9^8401 indicates a number of nines, thus: $0.9^8401 = 0.999\,999\,994\,01$, and $0.9^3032 = 0.999\,032$.

Proportional parts have not been given in this region because they would not be of sufficient accuracy.

EXAMPLES: $\Phi(1.2) = \text{Prob}\,(Z \leqslant 1.2) = 0.8849$; $\Phi(1.23) = 0.8907$; $\Phi(1.234) = 0.8914$; $\text{Prob}\,(Z \geqslant 2.3) = \Phi(-2.3) = 0.0107$ (making use of the symmetry of the normal distribution); $\text{Prob}\,(0.32 \leqslant Z \leqslant 1.43) = \Phi(1.43) - \Phi(0.32) = 0.9236 - 0.6255 = 0.2981$.

Other normal distributions may be dealt with by standardisation, i.e. by subtracting the mean and dividing by the standard deviation. For example if X has the normal distribution with mean 10.0 and standard deviation 2.0, $\text{Prob}\,(X \leqslant 17.5) = \text{Prob}\,(Z \leqslant \frac{1}{2}(17.5 - 10.0)) = \text{Prob}\,(Z \leqslant 3.75) = \Phi(3.75) = 0.9^4116 = 0.999\,911\,6$.

Percentage points of the normal distribution

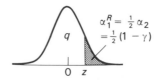

$q=\Phi(z)$	α_1^R	α_2	γ	z
0.50				0.0000
0.60	40%			0.2533
0.70	30%			0.5244
0.80	20%	40%	60%	0.8416
0.85	15%	30%	70%	1.0364
0.90	10%	20%	80%	1.2816
0.91	9%	18%	82%	1.3408
0.92	8%	16%	84%	1.4051
0.93	7%	14%	86%	1.4758
0.94	6%	12%	88%	1.5548
0.950	5.0%	10.0%	90.0%	1.6449
0.952	4.8%	9.6%	90.4%	1.6646
0.954	4.6%	9.2%	90.8%	1.6849
0.956	4.4%	8.8%	91.2%	1.7060
0.958	4.2%	8.4%	91.6%	1.7279
0.960	4.0%	8.0%	92.0%	1.7507
0.962	3.8%	7.6%	92.4%	1.7744
0.964	3.6%	7.2%	92.8%	1.7991
0.966	3.4%	6.8%	93.2%	1.8250
0.968	3.2%	6.4%	93.6%	1.8522
0.970	3.0%	6.0%	94.0%	1.8808
0.971	2.9%	5.8%	94.2%	1.8957
0.972	2.8%	5.6%	94.4%	1.9110
0.973	2.7%	5.4%	94.6%	1.9268
0.974	2.6%	5.2%	94.8%	1.9431
0.975	2.5%	5.0%	95.0%	1.9600
0.976	2.4%	4.8%	95.2%	1.9774
0.977	2.3%	4.6%	95.4%	1.9954
0.978	2.2%	4.4%	95.6%	2.0141
0.979	2.1%	4.2%	95.8%	2.0335
0.980	2.0%	4.0%	96.0%	2.0537
0.981	1.9%	3.8%	96.2%	2.0749
0.982	1.8%	3.6%	96.4%	2.0969
0.983	1.7%	3.4%	96.6%	2.1201
0.984	1.6%	3.2%	96.8%	2.1444
0.985	1.5%	3.0%	97.0%	2.1701
0.986	1.4%	2.8%	97.2%	2.1973
0.987	1.3%	2.6%	97.4%	2.2262
0.988	1.2%	2.4%	97.6%	2.2571
0.989	1.1%	2.2%	97.8%	2.2904
0.990	1.0%	2.0%	98.0%	2.3263
0.991	0.9%	1.8%	98.2%	2.3656
0.992	0.8%	1.6%	98.4%	2.4089
0.993	0.7%	1.4%	98.6%	2.4573
0.994	0.6%	1.2%	98.8%	2.5121
0.995	0.5%	1.0%	99.0%	2.5758
0.996	0.4%	0.8%	99.2%	2.6521
0.997	0.3%	0.6%	99.4%	2.7478
0.998	0.2%	0.4%	99.6%	2.8782
0.999	0.1%	0.2%	99.8%	3.0902
0.9995	0.05%	0.1%	99.9%	3.2905
0.9999	0.01%	0.02%	99.98%	3.7190
0.99995	0.005%	0.01%	99.99%	3.8906
0.99999	0.001%	0.002%	99.998%	4.2649
0.999995	0.0005%	0.001%	99.999%	4.4172
0.999999	0.0001%	0.0002%	99.9998%	4.7534
0.9999995	0.00005%	0.0001%	99.9999%	4.8916
0.9999999	0.00001%	0.00002%	99.99998%	5.1993
0.99999995	0.000005%	0.00001%	99.99999%	5.3267
0.99999999	0.000001%	0.000002%	99.999998%	5.6120

The following notation is used in this and subsequent tables. q represents a quantile, i.e. q and the tabulated value z are related here by $\mathrm{Prob}(Z\leqslant z)=q=\Phi(z)$; e.g. $\Phi(1.9600)=q=0.975$, where $z=1.9600$. α_1, α_1^L and α_1^R denote significance levels for one-tailed or one-sided critical regions. Sometimes α_1^L and α_1^R values, corresponding to critical regions in the left-hand and right-hand tails, need to be tabulated separately; in other cases one may easily be obtained from the other. Here we have included only α_1^R, since α_1^L values are obtained using the symmetry of the normal distribution. Thus if a 5% critical region in the right-hand tail is required, we find the entry corresponding to $\alpha_1^R=5\%$ and obtain $Z\geqslant1.6449$. Had we required a 5% critical region in the left-hand tail it would have been $Z\leqslant-1.6449$. α_2 gives critical regions for two-sided tests; here $|Z|\geqslant1.9600$ is the critical region for the two-sided test at the $\alpha_2=5\%$ significance level. Finally, γ indicates confidence levels for confidence intervals – so a 95% confidence interval here is derived from $|Z|\leqslant1.9600$. For example with a large sample $X_1,X_2,\ldots,X_n$ we know that $(\bar{X}-\mu)/(s/\sqrt{n})$ has approximately a standard normal distribution, where $\bar{X}=\Sigma X_i/n$ and the adjusted sample standard deviation s is given by $s=\{\Sigma(X_i-\bar{X})^2/(n-1)\}^{1/2}$. So a 95% confidence interval for μ is derived from $|(\bar{X}-\mu)/(s/\sqrt{n})|\leqslant1.9600$, which is equivalent to $\bar{X}-1.96s/\sqrt{n}\leqslant\mu\leqslant\bar{X}+1.96s/\sqrt{n}$.

Percentage points of the Student t distribution

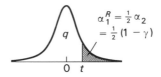

q	0.95	0.975	0.99	0.995
α_1^R	5%	2½%	1%	½%
α_2	10%	5%	2%	1%
γ	90%	95%	98%	99%
ν				
1	6.3138	12.7062	31.8205	63.6567
2	2.9200	4.3027	6.9646	9.9248
3	2.3534	3.1824	4.5407	5.8409
4	2.1318	2.7764	3.7469	4.6041
5	2.0150	2.5706	3.3649	4.0321
6	1.9432	2.4469	3.1427	3.7074
7	1.8946	2.3646	2.9980	3.4995
8	1.8595	2.3060	2.8965	3.3554
9	1.8331	2.2622	2.8214	3.2498
10	1.8125	2.2281	2.7638	3.1693
11	1.7959	2.2010	2.7181	3.1058
12	1.7823	2.1788	2.6810	3.0545
13	1.7709	2.1604	2.6503	3.0123
14	1.7613	2.1448	2.6245	2.9768
15	1.7531	2.1314	2.6025	2.9467
16	1.7459	2.1199	2.5835	2.9208
17	1.7396	2.1098	2.5669	2.8982
18	1.7341	2.1009	2.5524	2.8784
19	1.7291	2.0930	2.5395	2.8609
20	1.7247	2.0860	2.5280	2.8453
21	1.7207	2.0796	2.5176	2.8314
22	1.7171	2.0739	2.5083	2.8188
23	1.7139	2.0687	2.4999	2.8073
24	1.7109	2.0639	2.4922	2.7969
25	1.7081	2.0595	2.4851	2.7874
26	1.7056	2.0555	2.4786	2.7787
27	1.7033	2.0518	2.4727	2.7707
28	1.7011	2.0484	2.4671	2.7633
29	1.6991	2.0452	2.4620	2.7564
30	1.6973	2.0423	2.4573	2.7500
31	1.6955	2.0395	2.4528	2.7440
32	1.6939	2.0369	2.4487	2.7385
33	1.6924	2.0345	2.4448	2.7333
34	1.6909	2.0322	2.4411	2.7284
35	1.6896	2.0301	2.4377	2.7238
36	1.6883	2.0281	2.4345	2.7195
37	1.6871	2.0262	2.4314	2.7154
38	1.6860	2.0244	2.4286	2.7116
39	1.6849	2.0227	2.4258	2.7079
40	1.6839	2.0211	2.4233	2.7045
42	1.6820	2.0181	2.4185	2.6981
44	1.6802	2.0154	2.4141	2.6923
46	1.6787	2.0129	2.4102	2.6870
48	1.6772	2.0106	2.4066	2.6822
50	1.6759	2.0086	2.4033	2.6778
55	1.6730	2.0040	2.3961	2.6682
60	1.6706	2.0003	2.3901	2.6603
65	1.6686	1.9971	2.3851	2.6536
70	1.6669	1.9944	2.3808	2.6479
75	1.6654	1.9921	2.3771	2.6430
80	1.6641	1.9901	2.3739	2.6387
85	1.6630	1.9883	2.3710	2.6349
90	1.6620	1.9867	2.3685	2.6316
95	1.6611	1.9853	2.3662	2.6286
100	1.6602	1.9840	2.3642	2.6259
125	1.6571	1.9791	2.3565	2.6157
150	1.6551	1.9759	2.3515	2.6090
175	1.6536	1.9736	2.3478	2.6042
200	1.6525	1.9719	2.3451	2.6006
∞	1.6449	1.9600	2.3263	2.5758

The t distribution is mainly used for testing hypotheses and finding confidence intervals for means, given small samples from normal distributions. For a single sample, $(\bar{X}-\mu)/(s/\sqrt{n})$ has the t distribution with $\nu=n-1$ degrees of freedom (see notation above). So, e.g. if $n=10$, giving $\nu=9$, the $\gamma=95\%$ confidence interval for μ is $\bar{X}-2.2622s/\sqrt{10}\leqslant\mu\leqslant\bar{X}+2.2622s/\sqrt{10}$. Given two samples of sizes n_1 and n_2, sample means $\bar{X}_1$ and $\bar{X}_2$, and adjusted sample standard deviations s_1 and s_2, $(\bar{X}_1-\bar{X}_2)/\{s\sqrt{(1/n_1)+(1/n_2)}\}$ has the t distribution with $\nu=n_1+n_2-2$ degrees of freedom, where $s=[\{(n_1-1)s_1^2+(n_2-1)s_2^2\}/(n_1+n_2-2)]^{1/2}$. So if the population means are denoted μ_1 and μ_2, then to test $H_0:\mu_1=\mu_2$ against $H_1:\mu_1>\mu_2$ at the 5% level, given samples of sizes 6 and 10, the critical region is $(\bar{X}_1-\bar{X}_2)/(s\sqrt{\frac{1}{6}+\frac{1}{10}})\geqslant1.7613$, using $\nu=6+10-2=14$ and $\alpha_1^R=5\%$. As with the normal distribution, symmetry shows that α_1^L values are just the α_1^R values prefixed with a minus sign.

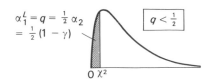
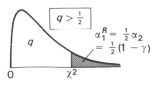

Percentage points of the chi-squared (χ^2) distribution

q	0.005	0.01	0.025	0.05	0.10	0.50	0.90	0.95	0.975	0.99	0.995
α_1^L	½%	1%	2½%	5%	10%						
α_1^R							10%	5%	2½%	1%	½%
α_2	1%	2%	5%	10%	20%		20%	10%	5%	2%	1%
γ	99%	98%	95%	90%	80%		80%	90%	95%	98%	99%
ν											
1	.00004	.00016	.00098	.00393	.0158	0.455	2.706	3.841	5.024	6.635	7.879
2	.0100	.0201	.0506	0.103	0.211	1.386	4.605	5.991	7.378	9.210	10.597
3	.0717	0.115	0.216	0.352	0.584	2.366	6.251	7.815	9.348	11.345	12.838
4	0.207	0.297	0.484	0.711	1.064	3.357	7.779	9.488	11.143	13.277	14.860
5	0.412	0.554	0.831	1.145	1.610	4.351	9.236	11.070	12.833	15.086	16.750
6	0.676	0.872	1.237	1.635	2.204	5.348	10.645	12.592	14.449	16.812	18.548
7	0.989	1.239	1.690	2.167	2.833	6.346	12.017	14.067	16.013	18.475	20.278
8	1.344	1.646	2.180	2.733	3.490	7.344	13.362	15.507	17.535	20.090	21.955
9	1.735	2.088	2.700	3.325	4.168	8.343	14.684	16.919	19.023	21.666	23.589
10	2.156	2.558	3.247	3.940	4.865	9.342	15.987	18.307	20.483	23.209	25.188
11	2.603	3.053	3.816	4.575	5.578	10.341	17.275	19.675	21.920	24.725	26.757
12	3.074	3.571	4.404	5.226	6.304	11.340	18.549	21.026	23.337	26.217	28.300
13	3.565	4.107	5.009	5.892	7.042	12.340	19.812	22.362	24.736	27.688	29.819
14	4.075	4.660	5.629	6.571	7.790	13.339	21.064	23.685	26.119	29.141	31.319
15	4.601	5.229	6.262	7.261	8.547	14.339	22.307	24.996	27.488	30.578	32.801
16	5.142	5.812	6.908	7.962	9.312	15.338	23.542	26.296	28.845	32.000	34.267
17	5.697	6.408	7.564	8.672	10.085	16.338	24.769	27.587	30.191	33.409	35.718
18	6.265	7.015	8.231	9.390	10.865	17.338	25.989	28.869	31.526	34.805	37.156
19	6.844	7.633	8.907	10.117	11.651	18.338	27.204	30.144	32.852	36.191	38.582
20	7.434	8.260	9.591	10.851	12.443	19.337	28.412	31.410	34.170	37.566	39.997
21	8.034	8.897	10.283	11.591	13.240	20.337	29.615	32.671	35.479	38.932	41.401
22	8.643	9.542	10.982	12.338	14.041	21.337	30.813	33.924	36.781	40.289	42.796
23	9.260	10.196	11.689	13.091	14.848	22.337	32.007	35.172	38.076	41.638	44.181
24	9.886	10.856	12.401	13.848	15.659	23.337	33.196	36.415	39.364	42.980	45.559
25	10.520	11.524	13.120	14.611	16.473	24.337	34.382	37.652	40.646	44.314	46.928
26	11.160	12.198	13.844	15.379	17.292	25.336	35.563	38.885	41.923	45.642	48.290
27	11.808	12.879	14.573	16.151	18.114	26.336	36.741	40.113	43.195	46.963	49.645
28	12.461	13.565	15.308	16.928	18.939	27.336	37.916	41.337	44.461	48.278	50.993
29	13.121	14.256	16.047	17.708	19.768	28.336	39.087	42.557	45.722	49.588	52.336
30	13.787	14.953	16.791	18.493	20.599	29.336	40.256	43.773	46.979	50.892	53.672
31	14.458	15.655	17.539	19.281	21.434	30.336	41.422	44.985	48.232	52.191	55.003
32	15.134	16.362	18.291	20.072	22.271	31.336	42.585	46.194	49.480	53.486	56.328
33	15.815	17.074	19.047	20.867	23.110	32.336	43.745	47.400	50.725	54.776	57.648
34	16.501	17.789	19.806	21.664	23.952	33.336	44.903	48.602	51.966	56.061	58.964
35	17.192	18.509	20.569	22.465	24.797	34.336	46.059	49.802	53.203	57.342	60.275
36	17.887	19.233	21.336	23.269	25.643	35.336	47.212	50.998	54.437	58.619	61.581
37	18.586	19.960	22.106	24.075	26.492	36.336	48.363	52.192	55.668	59.893	62.883
38	19.289	20.691	22.878	24.884	27.343	37.335	49.513	53.384	56.896	61.162	64.181
39	19.996	21.426	23.654	25.695	28.196	38.335	50.660	54.572	58.120	62.428	65.476
40	20.707	22.164	24.433	26.509	29.051	39.335	51.805	55.758	59.342	63.691	66.766
45	24.311	25.901	28.366	30.612	33.350	44.335	57.505	61.656	65.410	69.957	73.166
50	27.991	29.707	32.357	34.764	37.689	49.335	63.167	67.505	71.420	76.154	79.490
60	35.534	37.485	40.482	43.188	46.459	59.335	74.397	79.082	83.298	88.379	91.952
70	43.275	45.442	48.758	51.739	55.329	69.334	85.527	90.531	95.023	100.43	104.21
80	51.172	53.540	57.153	60.391	64.278	79.334	96.578	101.88	106.63	112.33	116.32
90	59.196	61.754	65.647	69.126	73.291	89.334	107.57	113.15	118.14	124.12	128.30
100	67.328	70.065	74.222	77.929	82.358	99.334	118.50	124.34	129.56	135.81	140.17
120	83.852	86.923	91.573	95.705	100.62	119.33	140.23	146.57	152.21	158.95	163.65
150	109.14	112.67	117.98	122.69	128.28	149.33	172.58	179.58	185.80	193.21	198.36
200	152.24	156.43	162.73	168.28	174.84	199.33	226.02	233.99	241.06	249.45	255.26

The χ^2 (chi-squared) distribution is used in testing hypotheses and forming confidence intervals for the standard deviation σ and the variance σ^2 of a normal population. Given a random sample of size n, $\chi^2 = (n-1)s^2/\sigma^2$ has the chi-squared distribution with $\nu = n-1$ degrees of freedom (s is defined on page 20). So if $n = 10$, giving $\nu = 9$, and the null hypothesis H_0 is $\sigma = 5$, 5% critical regions for testing against (a) H_1: $\sigma < 5$, (b) H_1: $\sigma > 5$ and (c) H_1: $\sigma \neq 5$ are (a) $9s^2/25 \leqslant 3.325$, (b) $9s^2/25 \geqslant 16.919$ and (c) $9s^2/25 \leqslant 2.700$ or $9s^2/25 \geqslant 19.023$, using significance levels (a) α_1^L, (b) α_1^R and (c) α_2 as appropriate. For example if $s^2 = 50.0$, this would result in rejection of H_0 in favour of H_1 at the 5% significance level in case (b) only. A $\gamma = 95\%$ confidence interval for σ with these data is derived from $2.700 \leqslant (n-1)s^2/\sigma^2 \leqslant 19.023$, i.e. $2.700 \leqslant 450.0/\sigma^2 \leqslant 19.023$, which gives $450/19.023 \leqslant \sigma^2 \leqslant 450/2.700$ or, taking square roots, $4.864 \leqslant \sigma \leqslant 12.910$.

The χ^2 distribution also gives critical values for the familiar χ^2 goodness-of-fit tests and tests for association in contingency tables (cross-tabulations). A classification scheme is given such that any observation must fall into precisely one class. The data then consist of frequency-counts and the statistic used is $\chi^2 = \Sigma (Ob. - Ex.)^2/Ex.$, where the sum is over all the classes, $Ob.$ denoting Observed frequencies and $Ex.$ Expected frequencies, these being calculated from the appropriate null hypothesis H_0. It is common to require that no expected frequencies be less than 5, and to regroup if necessary to achieve this. In goodness-of-fit tests, H_0 directly or indirectly specifies the probabilities of a random observation falling in each class. It is sometimes necessary to estimate population parameters (e.g. the mean and/or the standard deviation) to do this. The expected frequencies are these probabilities multiplied by the sample size. The number of degrees of freedom $\nu = $ (the number of classes $-1-$ the number of population parameters which have to be estimated). With contingency tables, H_0 is the hypothesis of no association between the classification schemes by rows and by columns, the expected frequency in any cell is (its row's subtotal) $\times$ (its column's subtotal) $\div$ (total number of observations), and the number of degrees of freedom ν is (number of rows -1) $\times$ (number of columns -1).

In all these cases, it is *large* values of χ^2 which are significant, so critical regions are of the form $\chi^2 \geqslant tabulated$ *value*, using α_1^R significance levels.

Percentage points of the F distribution

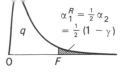

$$\alpha_1^R = \tfrac{1}{2}\alpha_2 = \tfrac{1}{2}(1-\gamma)$$

Three of the main uses of the F distribution are (a) the comparison of two variances, (b) to give critical values in the wide range of analysis-of-variance tests and (c) to find critical values for the multiple correlation coefficient.

(a) Comparison of two variances

Given random samples of sizes n_1 and n_2 from two normal populations having standard deviations σ_1 and σ_2 respectively, and where s_1 and s_2 denote the adjusted sample standard deviations (see page 20), $(s_1^2/s_2^2)/(\sigma_1^2/\sigma_2^2)$ has the F distribution with $(\nu_1, \nu_2) = (n_1 - 1, n_2 - 1)$ degrees of freedom. In the tables the degrees of freedom are given along the top (ν_1) and down the left-hand side (ν_2). For economy of space, the tables only give values in the right-hand tail of the distribution. This gives rise to minor inconvenience in some applications, which will be seen in the following illustrations:

(i) *One-sided test* — H_0: $\sigma_1 = \sigma_2$, H_1: $\sigma_1 > \sigma_2$.   The tabulated figures are directly appropriate. Thus if $n_1 = 5$ and $n_2 = 8$, giving $\nu_1 = 4$ and $\nu_2 = 7$, the $\alpha_1^R = 5\%$ critical region is $s_1^2/s_2^2 \geqslant 4.120$.

(ii) *One-sided test* – H_0: $\sigma_1 = \sigma_2$, H_1: $\sigma_1 < \sigma_2$.   Here we would normally need α_1^L values for s_1^2/s_2^2. However the tabulated values are appropriate if we use the statistic s_2^2/s_1^2 and switch round the degrees of freedom. So if $n_1 = 5$ and $n_2 = 8$, the appropriate $\alpha_1^R = 5\%$ critical region is $s_2^2/s_1^2 \geqslant 6.094$ (using $\nu_1 = 7$, $\nu_2 = 4$).

(iii) *Two-sided test* – H_0: $\sigma_1 = \sigma_2$, H_1: $\sigma_1 \neq \sigma_2$.   Calculate either s_1^2/s_2^2 or s_2^2/s_1^2, whichever is the larger, switching round the degrees of freedom if s_2^2/s_1^2 is chosen, and enter the tables using the α_2 significance levels. So if $n_1 = 5$ and $n_2 = 8$, giving $\nu_1 = 4$ and $\nu_2 = 7$, then we reject H_0 in favour of H_1 at the $\alpha_2 = 5\%$ significance level if either $s_1^2/s_2^2 \geqslant 5.523$ or $s_2^2/s_1^2 \geqslant 9.074$.

(iv) *Confidence interval for* σ_1/σ_2 *or* σ_1^2/σ_2^2.   This is derived from an interval of the form $f_1 \leqslant (s_1^2/s_2^2)/(\sigma_1^2/\sigma_2^2) \leqslant f_2$ where f_2 is read directly from the tables, using the desired confidence level γ, and f_1 is the reciprocal of the tabulated value found after switching the degrees of freedom. Thus if $\gamma = 95\%$, and $n_1 = 5$, $n_2 = 8$ giving $\nu_1 = 4$, $\nu_2 = 7$ again, then $f_2 = 5.523$ and $f_1 = 1/9.074$. So, e.g. if $s_1^2/s_2^2 = 4.0$ we

have $1/9.074 \leqslant 4.0/(\sigma_1^2/\sigma_2^2) \leqslant 5.523$ which, after a little manipulation, gives $4.0/5.523 \leqslant \sigma_1^2/\sigma_2^2 \leqslant 4.0 \times 9.074$, and taking square roots yields $(0.851 : 6.025)$ as the $\gamma = 95\%$ confidence interval for σ_1/σ_2.

(b) Analysis-of-variance (ANOVA) tests

The F statistics produced in the standard analysis-of-variance procedures are in the correct form for direct application of the tables, i.e. the critical regions are $F \geqslant tabulated\ value$. *Note that α_1^R (not α_2) significance levels should be used.* In the one-way classification analysis-of-variance, ν_1 is one less than the number of samples being compared; otherwise in experiments where more than one factor is involved, F statistics can be found to test the effect of each of the factors and ν_1 is then one less than the number of levels of the particular factor being examined. If an F statistic is being used to test for an interactive effect between two or more factors, ν_1 is the product of the numbers of degrees of freedom for the component factors. ν_2 is the number of degrees of freedom in the residual (or error, or within-sample) sum of squares, and is usually calculated as (total number of observations -1) $-$ (total number of degrees of freedom attributable to individual factors and their interactions (if relevant)). If the experiment includes replication, and a replication effect is included in the underlying model, this also counts as a factor for these purposes.

(c) Testing a multiple correlation coefficient

In a multiple linear regression $\hat{Y} = a_0 + a_1 X_1 + a_2 X_2 + \ldots + a_k X_k$, where $a_0, a_1, a_2, \ldots, a_k$ are estimated by least squares, the multiple correlation coefficient R is a measure of the goodness-of-fit of the regression model. R can be calculated as $R = +\sqrt{\Sigma(\hat{Y} - \bar{Y})^2/\Sigma(Y - \bar{Y})^2}$, where Y denotes the observed values and $\bar{Y}$ their mean. R is also the linear correlation coefficient of $\hat{Y}$ with Y. Assuming normality of residuals, R can be used to test if the regression model is useful. Calculate $F = (n - k - 1)R^2/k(1 - R^2)$, where n is the size of the sample from which R was computed, and the critical regions showing evidence that the model is indeed useful are of the form $F \geqslant tabulated\ value$, using the F tables with $\nu_1 = k$, $\nu_2 = n - k - 1$ and α_1^R significance levels.

		$q = 0.90$		$\alpha_1^R = 10\%$		$\alpha_2 = 20\%$		$\gamma = 80\%$	

ν_2 \ ν_1	1	2	3	4	5	6	7	8	9	10	12	15	20	25	30	50	75	100	150	∞	ν_2
1	39.86	49.50	53.59	55.83	57.24	58.20	58.91	59.44	59.86	60.19	60.71	61.22	61.74	62.05	62.26	62.69	62.90	63.01	63.11	63.33	1
2	8.526	9.000	9.162	9.243	9.293	9.326	9.349	9.367	9.381	9.392	9.408	9.425	9.441	9.451	9.458	9.471	9.478	9.481	9.485	9.491	2
3	5.538	5.462	5.391	5.343	5.309	5.285	5.266	5.252	5.240	5.230	5.216	5.200	5.184	5.175	5.168	5.155	5.148	5.144	5.141	5.134	3
4	4.545	4.325	4.191	4.107	4.051	4.010	3.979	3.955	3.936	3.920	3.896	3.870	3.844	3.828	3.817	3.795	3.784	3.778	3.772	3.761	4
5	4.060	3.780	3.619	3.520	3.453	3.405	3.368	3.339	3.316	3.297	3.268	3.238	3.207	3.187	3.174	3.147	3.133	3.126	3.119	3.105	5
6	3.776	3.463	3.289	3.181	3.108	3.055	3.014	2.983	2.958	2.937	2.905	2.871	2.836	2.815	2.800	2.770	2.754	2.746	2.738	2.722	6
7	3.589	3.257	3.074	2.961	2.883	2.827	2.785	2.752	2.725	2.703	2.668	2.632	2.595	2.571	2.555	2.523	2.506	2.497	2.488	2.471	7
8	3.458	3.113	2.924	2.806	2.726	2.668	2.624	2.589	2.561	2.538	2.502	2.464	2.425	2.400	2.383	2.348	2.330	2.321	2.312	2.293	8
9	3.360	3.006	2.813	2.693	2.611	2.551	2.505	2.469	2.440	2.416	2.379	2.340	2.298	2.272	2.255	2.218	2.199	2.189	2.179	2.159	9
10	3.285	2.924	2.728	2.605	2.522	2.461	2.414	2.377	2.347	2.323	2.284	2.244	2.201	2.174	2.155	2.117	2.097	2.087	2.077	2.055	10
11	3.225	2.860	2.660	2.536	2.451	2.389	2.342	2.304	2.274	2.248	2.209	2.167	2.123	2.095	2.076	2.036	2.016	2.005	1.994	1.972	11
12	3.177	2.807	2.606	2.480	2.394	2.331	2.283	2.245	2.214	2.188	2.147	2.105	2.060	2.031	2.011	1.970	1.949	1.938	1.927	1.904	12
13	3.136	2.763	2.560	2.434	2.347	2.283	2.234	2.195	2.164	2.138	2.097	2.053	2.007	1.978	1.958	1.915	1.893	1.882	1.870	1.846	13
14	3.102	2.726	2.522	2.395	2.307	2.243	2.193	2.154	2.122	2.095	2.054	2.010	1.962	1.933	1.912	1.869	1.846	1.834	1.822	1.797	14
15	3.073	2.695	2.490	2.361	2.273	2.208	2.158	2.119	2.086	2.059	2.017	1.972	1.924	1.894	1.873	1.828	1.805	1.793	1.781	1.755	15
16	3.048	2.668	2.462	2.333	2.244	2.178	2.128	2.088	2.055	2.028	1.985	1.940	1.891	1.860	1.839	1.793	1.769	1.757	1.744	1.718	16
17	3.026	2.645	2.437	2.308	2.218	2.152	2.102	2.061	2.028	2.001	1.958	1.912	1.862	1.831	1.809	1.763	1.738	1.726	1.713	1.686	17
18	3.007	2.624	2.416	2.286	2.196	2.130	2.079	2.038	2.005	1.977	1.933	1.887	1.837	1.805	1.783	1.736	1.711	1.698	1.684	1.657	18
19	2.990	2.606	2.397	2.266	2.176	2.109	2.058	2.017	1.984	1.956	1.912	1.865	1.814	1.782	1.759	1.711	1.686	1.673	1.659	1.631	19
20	2.975	2.589	2.380	2.249	2.158	2.091	2.040	1.999	1.965	1.937	1.892	1.845	1.794	1.761	1.738	1.690	1.664	1.650	1.636	1.607	20
21	2.961	2.575	2.365	2.233	2.142	2.075	2.023	1.982	1.948	1.920	1.875	1.827	1.776	1.742	1.719	1.670	1.644	1.630	1.616	1.586	21
22	2.949	2.561	2.351	2.219	2.128	2.060	2.008	1.967	1.933	1.904	1.859	1.811	1.759	1.726	1.702	1.652	1.625	1.611	1.597	1.567	22
23	2.937	2.549	2.339	2.207	2.115	2.047	1.995	1.953	1.919	1.890	1.845	1.796	1.744	1.710	1.686	1.636	1.609	1.594	1.580	1.549	23
24	2.927	2.538	2.327	2.195	2.103	2.035	1.983	1.941	1.906	1.877	1.832	1.783	1.730	1.696	1.672	1.621	1.593	1.579	1.564	1.533	24
25	2.918	2.528	2.317	2.184	2.092	2.024	1.971	1.929	1.895	1.866	1.820	1.771	1.718	1.683	1.659	1.607	1.579	1.565	1.549	1.518	25
30	2.881	2.489	2.276	2.142	2.049	1.980	1.927	1.884	1.849	1.819	1.773	1.722	1.667	1.632	1.606	1.552	1.523	1.507	1.491	1.456	30
35	2.855	2.461	2.247	2.113	2.019	1.950	1.896	1.852	1.817	1.787	1.739	1.688	1.632	1.595	1.569	1.513	1.482	1.465	1.448	1.411	35
40	2.835	2.440	2.226	2.091	1.997	1.927	1.873	1.829	1.793	1.763	1.715	1.662	1.605	1.568	1.541	1.483	1.451	1.434	1.416	1.377	40
50	2.809	2.412	2.197	2.061	1.966	1.895	1.840	1.796	1.760	1.729	1.680	1.627	1.568	1.529	1.502	1.441	1.407	1.388	1.369	1.327	50
75	2.774	2.375	2.158	2.021	1.926	1.854	1.798	1.754	1.716	1.685	1.635	1.580	1.519	1.478	1.449	1.384	1.346	1.326	1.304	1.254	75
100	2.756	2.356	2.139	2.002	1.906	1.834	1.778	1.732	1.695	1.663	1.612	1.557	1.494	1.453	1.423	1.355	1.315	1.293	1.270	1.214	100
150	2.739	2.338	2.121	1.983	1.886	1.814	1.757	1.712	1.674	1.642	1.590	1.533	1.470	1.427	1.396	1.325	1.283	1.259	1.233	1.169	150
∞	2.706	2.303	2.084	1.945	1.847	1.774	1.717	1.670	1.632	1.599	1.546	1.487	1.421	1.375	1.342	1.263	1.214	1.185	1.151	(1.0)	∞

Percentage points of the F distribution

$q = 0.95$	$\alpha_1^R = 5\%$	$\alpha_2 = 10\%$	$\gamma = 90\%$

ν_1 ν_2	1	2	3	4	5	6	7	8	9	10	12	15	20	25	30	50	75	100	150	∞	ν_2
1	161.4	199.5	215.7	224.6	230.2	234.0	236.8	238.9	240.5	241.9	243.9	245.9	248.0	249.3	250.1	251.8	252.6	253.0	253.5	254.3	1
2	18.51	19.00	19.16	19.25	19.30	19.33	19.35	19.37	19.38	19.40	19.41	19.43	19.45	19.46	19.46	19.48	19.48	19.49	19.49	19.50	2
3	10.13	9.552	9.277	9.117	9.013	8.941	8.887	8.845	8.812	8.786	8.745	8.703	8.660	8.634	8.617	8.581	8.563	8.554	8.545	8.526	3
4	7.709	6.944	6.591	6.388	6.256	6.163	6.094	6.041	5.999	5.964	5.912	5.858	5.803	5.769	5.746	5.699	5.676	5.664	5.652	5.628	4
5	6.608	5.786	5.409	5.192	5.050	4.950	4.876	4.818	4.772	4.735	4.678	4.619	4.558	4.521	4.496	4.444	4.418	4.405	4.392	4.365	5
6	5.987	5.143	4.757	4.534	4.387	4.284	4.207	4.147	4.099	4.060	4.000	3.938	3.874	3.835	3.808	3.754	3.726	3.712	3.698	3.669	6
7	5.591	4.737	4.347	4.120	3.972	3.866	3.787	3.726	3.677	3.637	3.575	3.511	3.445	3.404	3.376	3.319	3.290	3.275	3.260	3.230	7
8	5.318	4.459	4.066	3.838	3.687	3.581	3.500	3.438	3.388	3.347	3.284	3.218	3.150	3.108	3.079	3.020	2.990	2.975	2.959	2.928	8
9	5.117	4.256	3.863	3.633	3.482	3.374	3.293	3.230	3.179	3.137	3.073	3.006	2.936	2.893	2.864	2.803	2.771	2.756	2.739	2.707	9
10	4.965	4.103	3.708	3.478	3.326	3.217	3.135	3.072	3.020	2.978	2.913	2.845	2.774	2.730	2.700	2.637	2.605	2.588	2.572	2.538	10
11	4.844	3.982	3.587	3.357	3.204	3.095	3.012	2.948	2.896	2.854	2.788	2.719	2.646	2.601	2.570	2.507	2.473	2.457	2.439	2.404	11
12	4.747	3.885	3.490	3.259	3.106	2.996	2.913	2.849	2.796	2.753	2.687	2.617	2.544	2.498	2.466	2.401	2.367	2.350	2.332	2.296	12
13	4.667	3.806	3.411	3.179	3.025	2.915	2.832	2.767	2.714	2.671	2.604	2.533	2.459	2.412	2.380	2.314	2.279	2.261	2.243	2.206	13
14	4.600	3.739	3.344	3.112	2.958	2.848	2.764	2.699	2.646	2.602	2.534	2.463	2.388	2.341	2.308	2.241	2.205	2.187	2.169	2.131	14
15	4.543	3.682	3.287	3.056	2.901	2.790	2.707	2.641	2.588	2.544	2.475	2.403	2.328	2.280	2.247	2.178	2.142	2.123	2.105	2.066	15
16	4.494	3.634	3.239	3.007	2.852	2.741	2.657	2.591	2.538	2.494	2.425	2.352	2.276	2.227	2.194	2.124	2.087	2.068	2.049	2.010	16
17	4.451	3.592	3.197	2.965	2.810	2.699	2.614	2.548	2.494	2.450	2.381	2.308	2.230	2.181	2.148	2.077	2.040	2.020	2.001	1.960	17
18	4.414	3.555	3.160	2.928	2.773	2.661	2.577	2.510	2.456	2.412	2.342	2.269	2.191	2.141	2.107	2.035	1.998	1.978	1.958	1.917	18
19	4.381	3.522	3.127	2.895	2.740	2.628	2.544	2.477	2.423	2.378	2.308	2.234	2.155	2.106	2.071	1.999	1.960	1.940	1.920	1.878	19
20	4.351	3.493	3.098	2.866	2.711	2.599	2.514	2.447	2.393	2.348	2.278	2.203	2.124	2.074	2.039	1.966	1.927	1.907	1.886	1.843	20
21	4.325	3.467	3.072	2.840	2.685	2.573	2.488	2.420	2.366	2.321	2.250	2.176	2.096	2.045	2.010	1.936	1.897	1.876	1.855	1.812	21
22	4.301	3.443	3.049	2.817	2.661	2.549	2.464	2.397	2.342	2.297	2.226	2.151	2.071	2.020	1.984	1.909	1.869	1.849	1.827	1.783	22
23	4.279	3.422	3.028	2.796	2.640	2.528	2.442	2.375	2.320	2.275	2.204	2.128	2.048	1.996	1.961	1.885	1.844	1.823	1.802	1.757	23
24	4.260	3.403	3.009	2.776	2.621	2.508	2.423	2.355	2.300	2.255	2.183	2.108	2.027	1.975	1.939	1.863	1.822	1.800	1.779	1.733	24
25	4.242	3.385	2.991	2.759	2.603	2.490	2.405	2.337	2.282	2.236	2.165	2.089	2.007	1.955	1.919	1.842	1.801	1.779	1.757	1.711	25
30	4.171	3.316	2.922	2.690	2.534	2.421	2.334	2.266	2.211	2.165	2.092	2.015	1.932	1.878	1.841	1.761	1.718	1.695	1.672	1.622	30
35	4.121	3.267	2.874	2.641	2.485	2.372	2.285	2.217	2.161	2.114	2.041	1.963	1.878	1.824	1.786	1.703	1.659	1.635	1.610	1.559	35
40	4.085	3.232	2.839	2.606	2.449	2.336	2.249	2.180	2.124	2.077	2.003	1.924	1.839	1.783	1.744	1.660	1.614	1.589	1.564	1.509	40
50	4.034	3.183	2.790	2.557	2.400	2.286	2.199	2.130	2.073	2.026	1.952	1.871	1.784	1.727	1.687	1.599	1.551	1.525	1.498	1.438	50
75	3.968	3.119	2.727	2.494	2.337	2.222	2.134	2.064	2.007	1.959	1.884	1.802	1.712	1.653	1.611	1.518	1.466	1.437	1.407	1.338	75
100	3.936	3.087	2.696	2.463	2.305	2.191	2.103	2.032	1.975	1.927	1.850	1.768	1.676	1.616	1.573	1.477	1.422	1.392	1.359	1.283	100
150	3.904	3.056	2.665	2.432	2.274	2.160	2.071	2.001	1.943	1.894	1.817	1.734	1.641	1.580	1.535	1.436	1.377	1.345	1.309	1.223	150
∞	3.841	2.996	2.605	2.372	2.214	2.099	2.010	1.938	1.880	1.831	1.752	1.666	1.571	1.506	1.459	1.350	1.283	1.243	1.197	(1.0)	∞

$q = 0.975$	$\alpha_1^R = 2\frac{1}{2}\%$	$\alpha_2 = 5\%$	$\gamma = 95\%$

ν_1 ν_2	1	2	3	4	5	6	7	8	9	10	12	15	20	25	30	50	75	100	150	∞	ν_2
1	647.8	799.5	864.2	899.6	921.8	937.1	948.2	956.7	963.3	968.6	976.7	984.9	993.1	998.1	1001	1008	1011	1013	1015	1018	1
2	38.51	39.00	39.17	39.25	39.30	39.33	39.36	39.37	39.39	39.40	39.41	39.43	39.45	39.46	39.46	39.48	39.48	39.49	39.49	39.50	2
3	17.44	16.04	15.44	15.10	14.88	14.73	14.62	14.54	14.47	14.42	14.34	14.25	14.17	14.12	14.08	14.01	13.97	13.96	13.94	13.90	3
4	12.22	10.65	9.979	9.605	9.364	9.197	9.074	8.980	8.905	8.844	8.751	8.657	8.560	8.501	8.461	8.381	8.340	8.319	8.299	8.257	4
5	10.01	8.434	7.764	7.388	7.146	6.978	6.853	6.757	6.681	6.619	6.525	6.428	6.329	6.268	6.227	6.144	6.101	6.080	6.059	6.015	5
6	8.813	7.260	6.599	6.227	5.988	5.820	5.695	5.600	5.523	5.461	5.366	5.269	5.168	5.107	5.065	4.980	4.937	4.915	4.893	4.849	6
7	8.073	6.542	5.890	5.523	5.285	5.119	4.995	4.899	4.823	4.761	4.666	4.568	4.467	4.405	4.362	4.276	4.232	4.210	4.188	4.142	7
8	7.571	6.059	5.416	5.053	4.817	4.652	4.529	4.433	4.357	4.295	4.200	4.101	3.999	3.937	3.894	3.807	3.762	3.739	3.716	3.670	8
9	7.209	5.715	5.078	4.718	4.484	4.320	4.197	4.102	4.026	3.964	3.868	3.769	3.667	3.604	3.560	3.472	3.426	3.403	3.380	3.333	9
10	6.937	5.456	4.826	4.468	4.236	4.072	3.950	3.855	3.779	3.717	3.621	3.522	3.419	3.355	3.311	3.221	3.175	3.152	3.128	3.080	10
11	6.724	5.256	4.630	4.275	4.044	3.881	3.759	3.664	3.588	3.526	3.430	3.330	3.226	3.162	3.118	3.027	2.980	2.956	2.932	2.883	11
12	6.554	5.096	4.474	4.121	3.891	3.728	3.607	3.512	3.436	3.374	3.277	3.177	3.073	3.008	2.963	2.871	2.824	2.800	2.775	2.725	12
13	6.414	4.965	4.347	3.996	3.767	3.604	3.483	3.388	3.312	3.250	3.153	3.053	2.948	2.882	2.837	2.744	2.696	2.671	2.647	2.595	13
14	6.298	4.857	4.242	3.892	3.663	3.501	3.380	3.285	3.209	3.147	3.050	2.949	2.844	2.778	2.732	2.638	2.590	2.565	2.539	2.487	14
15	6.200	4.765	4.153	3.804	3.576	3.415	3.293	3.199	3.123	3.060	2.963	2.862	2.756	2.689	2.644	2.549	2.499	2.474	2.448	2.395	15
16	6.115	4.687	4.077	3.729	3.502	3.341	3.219	3.125	3.049	2.986	2.889	2.788	2.681	2.614	2.568	2.472	2.422	2.396	2.370	2.316	16
17	6.042	4.619	4.011	3.665	3.438	3.277	3.156	3.061	2.985	2.922	2.825	2.723	2.616	2.548	2.502	2.405	2.355	2.329	2.302	2.247	17
18	5.978	4.560	3.954	3.608	3.382	3.221	3.100	3.005	2.929	2.866	2.769	2.667	2.559	2.491	2.445	2.347	2.296	2.269	2.242	2.187	18
19	5.922	4.508	3.903	3.559	3.333	3.172	3.051	2.956	2.880	2.817	2.720	2.617	2.509	2.441	2.394	2.295	2.243	2.217	2.190	2.133	19
20	5.871	4.461	3.859	3.515	3.289	3.128	3.007	2.913	2.837	2.774	2.676	2.573	2.464	2.396	2.349	2.249	2.197	2.170	2.142	2.085	20
21	5.827	4.420	3.819	3.475	3.250	3.090	2.969	2.874	2.798	2.735	2.637	2.534	2.425	2.356	2.308	2.208	2.155	2.128	2.100	2.042	21
22	5.786	4.383	3.783	3.440	3.215	3.055	2.934	2.839	2.763	2.700	2.602	2.498	2.389	2.320	2.272	2.171	2.118	2.090	2.062	2.003	22
23	5.750	4.349	3.750	3.408	3.183	3.023	2.902	2.808	2.731	2.668	2.570	2.466	2.357	2.287	2.239	2.137	2.084	2.056	2.027	1.968	23
24	5.717	4.319	3.721	3.379	3.155	2.995	2.874	2.779	2.703	2.640	2.541	2.437	2.327	2.257	2.209	2.107	2.052	2.024	1.995	1.935	24
25	5.686	4.291	3.694	3.353	3.129	2.969	2.848	2.753	2.677	2.613	2.515	2.411	2.300	2.230	2.182	2.079	2.024	1.996	1.966	1.906	25
30	5.568	4.182	3.589	3.250	3.026	2.867	2.746	2.651	2.575	2.511	2.412	2.307	2.195	2.124	2.074	1.968	1.911	1.882	1.851	1.787	30
35	5.485	4.106	3.517	3.179	2.956	2.796	2.676	2.581	2.504	2.440	2.341	2.235	2.122	2.049	1.999	1.890	1.832	1.801	1.769	1.702	35
40	5.424	4.051	3.463	3.126	2.904	2.744	2.624	2.529	2.452	2.388	2.288	2.182	2.068	1.994	1.943	1.832	1.772	1.741	1.700	1.637	40
50	5.340	3.975	3.390	3.054	2.833	2.674	2.553	2.458	2.381	2.317	2.216	2.109	1.993	1.919	1.866	1.752	1.689	1.656	1.621	1.545	50
75	5.232	3.876	3.296	2.962	2.741	2.582	2.461	2.366	2.289	2.224	2.123	2.014	1.896	1.819	1.765	1.645	1.578	1.542	1.503	1.417	75
100	5.179	3.828	3.250	2.917	2.696	2.537	2.417	2.321	2.244	2.179	2.077	1.968	1.849	1.770	1.715	1.592	1.522	1.483	1.442	1.347	100
150	5.126	3.781	3.204	2.872	2.652	2.494	2.373	2.278	2.200	2.135	2.032	1.922	1.801	1.722	1.665	1.538	1.464	1.423	1.379	1.271	150
∞	5.024	3.689	3.116	2.786	2.567	2.408	2.288	2.192	2.114	2.048	1.945	1.833	1.708	1.626	1.566	1.428	1.345	1.296	1.239	(1.0)	∞

23

Percentage points of the F distribution

	$q = 0.99$	$\alpha_1^R = 1\%$	$\alpha_2 = 2\%$	$\gamma = 98\%$

ν_2\ν_1	1	2	3	4	5	6	7	8	9	10	12	15	20	25	30	50	75	100	150	∞
1	4052	4999	5403	5625	5764	5859	5928	5981	6022	6056	6106	6157	6209	6240	6261	6303	6324	6334	6345	6366
2	98.50	99.00	99.17	99.25	99.30	99.33	99.36	99.37	99.39	99.40	99.42	99.43	99.45	99.46	99.47	99.48	99.49	99.49	99.49	99.50
3	34.12	30.82	29.46	28.71	28.24	27.91	27.67	27.49	27.35	27.23	27.05	26.87	26.69	26.58	26.50	26.35	26.28	26.24	26.20	26.13
4	21.20	18.00	16.69	15.98	15.52	15.21	14.98	14.80	14.66	14.55	14.37	14.20	14.02	13.91	13.84	13.69	13.61	13.58	13.54	13.46
5	16.26	13.27	12.06	11.39	10.97	10.67	10.46	10.29	10.16	10.05	9.888	9.722	9.553	9.449	9.379	9.238	9.166	9.130	9.094	9.020
6	13.75	10.92	9.780	9.148	8.746	8.466	8.260	8.102	7.976	7.874	7.718	7.559	7.396	7.296	7.229	7.091	7.022	6.987	6.951	6.880
7	12.25	9.547	8.451	7.847	7.460	7.191	6.993	6.840	6.719	6.620	6.469	6.314	6.155	6.058	5.992	5.858	5.789	5.755	5.720	5.650
8	11.26	8.649	7.591	7.006	6.632	6.371	6.178	6.029	5.911	5.814	5.667	5.515	5.359	5.263	5.198	5.065	4.998	4.963	4.929	4.859
9	10.56	8.022	6.992	6.422	6.057	5.802	5.613	5.467	5.351	5.257	5.111	4.962	4.808	4.713	4.649	4.517	4.449	4.415	4.380	4.311
10	10.04	7.559	6.552	5.994	5.636	5.386	5.200	5.057	4.942	4.849	4.706	4.558	4.405	4.311	4.247	4.115	4.048	4.014	3.979	3.909
11	9.646	7.206	6.217	5.668	5.316	5.069	4.886	4.744	4.632	4.539	4.397	4.251	4.099	4.005	3.941	3.810	3.742	3.708	3.673	3.602
12	9.330	6.927	5.953	5.412	5.064	4.821	4.640	4.499	4.388	4.296	4.155	4.010	3.858	3.765	3.701	3.569	3.501	3.467	3.432	3.361
13	9.074	6.701	5.739	5.205	4.862	4.620	4.441	4.302	4.191	4.100	3.960	3.815	3.665	3.571	3.507	3.375	3.307	3.272	3.237	3.165
14	8.862	6.515	5.564	5.035	4.695	4.456	4.278	4.140	4.030	3.939	3.800	3.656	3.505	3.412	3.348	3.215	3.147	3.112	3.076	3.004
15	8.683	6.359	5.417	4.893	4.556	4.318	4.142	4.004	3.895	3.805	3.666	3.522	3.372	3.278	3.214	3.081	3.012	2.977	2.942	2.868
16	8.531	6.226	5.292	4.773	4.437	4.202	4.026	3.890	3.780	3.691	3.553	3.409	3.259	3.165	3.101	2.967	2.898	2.863	2.827	2.753
17	8.400	6.112	5.185	4.669	4.336	4.102	3.927	3.791	3.682	3.593	3.455	3.312	3.162	3.068	3.003	2.869	2.800	2.764	2.728	2.653
18	8.285	6.013	5.092	4.579	4.248	4.015	3.841	3.705	3.597	3.508	3.371	3.227	3.077	2.983	2.919	2.784	2.714	2.678	2.641	2.566
19	8.185	5.926	5.010	4.500	4.171	3.939	3.765	3.631	3.523	3.434	3.297	3.153	3.003	2.909	2.844	2.709	2.639	2.602	2.565	2.489
20	8.096	5.849	4.938	4.431	4.103	3.871	3.699	3.564	3.457	3.368	3.231	3.088	2.938	2.843	2.778	2.643	2.572	2.535	2.498	2.421
21	8.017	5.780	4.874	4.369	4.042	3.812	3.640	3.506	3.398	3.310	3.173	3.030	2.880	2.785	2.720	2.584	2.512	2.475	2.438	2.360
22	7.945	5.719	4.817	4.313	3.988	3.758	3.587	3.453	3.346	3.258	3.121	2.978	2.827	2.733	2.667	2.531	2.459	2.422	2.384	2.305
23	7.881	5.664	4.765	4.264	3.939	3.710	3.539	3.406	3.299	3.211	3.074	2.931	2.781	2.686	2.620	2.483	2.411	2.373	2.335	2.256
24	7.823	5.614	4.718	4.218	3.895	3.667	3.496	3.363	3.256	3.168	3.032	2.889	2.738	2.643	2.577	2.440	2.367	2.329	2.291	2.211
25	7.770	5.568	4.675	4.177	3.855	3.627	3.457	3.324	3.217	3.129	2.993	2.850	2.699	2.604	2.538	2.400	2.327	2.289	2.250	2.169
30	7.562	5.390	4.510	4.018	3.699	3.473	3.304	3.173	3.067	2.979	2.843	2.700	2.549	2.453	2.386	2.245	2.170	2.131	2.091	2.006
35	7.419	5.268	4.396	3.908	3.592	3.368	3.200	3.069	2.963	2.876	2.740	2.597	2.445	2.348	2.281	2.137	2.060	2.020	1.979	1.891
40	7.314	5.179	4.313	3.828	3.514	3.291	3.124	2.993	2.888	2.801	2.665	2.522	2.369	2.271	2.203	2.058	1.980	1.938	1.896	1.805
50	7.171	5.057	4.199	3.720	3.408	3.186	3.020	2.890	2.785	2.698	2.562	2.419	2.265	2.167	2.098	1.949	1.868	1.825	1.780	1.683
75	6.985	4.900	4.054	3.580	3.272	3.052	2.887	2.758	2.653	2.567	2.431	2.287	2.132	2.031	1.960	1.806	1.720	1.674	1.625	1.516
100	6.895	4.824	3.984	3.513	3.206	2.988	2.823	2.694	2.590	2.503	2.368	2.223	2.067	1.965	1.893	1.735	1.646	1.598	1.546	1.427
150	6.807	4.749	3.915	3.447	3.142	2.924	2.761	2.632	2.528	2.441	2.305	2.160	2.003	1.900	1.827	1.665	1.572	1.520	1.465	1.331
∞	6.635	4.605	3.782	3.319	3.017	2.802	2.639	2.511	2.407	2.321	2.185	2.039	1.878	1.773	1.696	1.523	1.419	1.358	1.288	(1.0)

	$q = 0.995$	$\alpha_1^R = \frac{1}{2}\%$	$\alpha_2 = 1\%$	$\gamma = 99\%$

ν_2\ν_1	1	2	3	4	5	6	7	8	9	10	12	15	20	25	30	50	75	100	150	∞
1	16211	20000	21615	22500	23056	23437	23715	23925	24091	24224	24426	24630	24836	24960	25044	25211	25295	25337	25380	25464
2	198.5	199.0	199.2	199.2	199.3	199.3	199.4	199.4	199.4	199.4	199.4	199.4	199.4	199.5	199.5	199.5	199.5	199.5	199.5	199.5
3	55.55	49.80	47.47	46.19	45.39	44.84	44.43	44.13	43.88	43.69	43.39	43.08	42.78	42.59	42.47	42.21	42.09	42.02	41.96	41.83
4	31.33	26.28	24.26	23.15	22.46	21.97	21.62	21.35	21.14	20.97	20.70	20.44	20.17	20.00	19.89	19.67	19.55	19.50	19.44	19.32
5	22.78	18.31	16.53	15.56	14.94	14.51	14.20	13.96	13.77	13.62	13.38	13.15	12.90	12.76	12.66	12.45	12.35	12.30	12.25	12.14
6	18.63	14.54	12.92	12.03	11.46	11.07	10.79	10.57	10.39	10.25	10.03	9.814	9.589	9.451	9.358	9.170	9.074	9.026	8.977	8.879
7	16.24	12.40	10.88	10.05	9.522	9.155	8.885	8.678	8.514	8.380	8.176	7.968	7.754	7.623	7.534	7.354	7.263	7.217	7.170	7.076
8	14.69	11.04	9.596	8.805	8.302	7.952	7.694	7.496	7.339	7.211	7.015	6.814	6.608	6.482	6.396	6.222	6.133	6.088	6.042	5.951
9	13.61	10.11	8.717	7.956	7.471	7.134	6.885	6.693	6.541	6.417	6.227	6.032	5.832	5.708	5.625	5.454	5.367	5.322	5.278	5.188
10	12.83	9.427	8.081	7.343	6.872	6.545	6.302	6.116	5.968	5.847	5.661	5.471	5.274	5.153	5.071	4.902	4.816	4.772	4.728	4.639
11	12.23	8.912	7.600	6.881	6.422	6.102	5.865	5.682	5.537	5.418	5.236	5.049	4.855	4.736	4.654	4.488	4.402	4.359	4.315	4.226
12	11.75	8.510	7.226	6.521	6.071	5.757	5.525	5.345	5.202	5.085	4.906	4.721	4.530	4.412	4.331	4.165	4.080	4.037	3.993	3.904
13	11.37	8.186	6.926	6.233	5.791	5.482	5.253	5.076	4.935	4.820	4.643	4.460	4.270	4.153	4.073	3.908	3.823	3.780	3.736	3.647
14	11.06	7.922	6.680	5.998	5.562	5.257	5.031	4.857	4.717	4.603	4.428	4.247	4.059	3.942	3.862	3.698	3.612	3.569	3.525	3.436
15	10.80	7.701	6.476	5.803	5.372	5.071	4.847	4.674	4.536	4.424	4.250	4.070	3.883	3.766	3.687	3.523	3.437	3.394	3.350	3.260
16	10.58	7.514	6.303	5.638	5.212	4.913	4.692	4.521	4.384	4.272	4.099	3.920	3.734	3.618	3.539	3.375	3.290	3.246	3.202	3.112
17	10.38	7.354	6.156	5.497	5.075	4.779	4.559	4.389	4.254	4.142	3.971	3.793	3.607	3.492	3.412	3.248	3.163	3.119	3.075	2.984
18	10.22	7.215	6.028	5.375	4.956	4.663	4.445	4.276	4.141	4.030	3.860	3.683	3.498	3.382	3.303	3.139	3.053	3.009	2.965	2.873
19	10.07	7.093	5.916	5.268	4.853	4.561	4.345	4.177	4.043	3.933	3.763	3.587	3.402	3.287	3.208	3.043	2.957	2.913	2.868	2.776
20	9.944	6.986	5.818	5.174	4.762	4.472	4.257	4.090	3.956	3.847	3.678	3.502	3.318	3.203	3.123	2.959	2.872	2.828	2.783	2.690
21	9.830	6.891	5.730	5.091	4.681	4.393	4.179	4.013	3.880	3.771	3.602	3.427	3.243	3.128	3.049	2.884	2.797	2.753	2.707	2.614
22	9.727	6.806	5.652	5.017	4.609	4.322	4.109	3.944	3.812	3.703	3.535	3.360	3.176	3.061	2.982	2.817	2.730	2.685	2.640	2.545
23	9.635	6.730	5.582	4.950	4.544	4.259	4.047	3.882	3.750	3.642	3.475	3.300	3.116	3.001	2.922	2.756	2.669	2.624	2.579	2.484
24	9.551	6.661	5.519	4.890	4.486	4.202	3.991	3.826	3.695	3.587	3.420	3.246	3.062	2.947	2.868	2.702	2.614	2.569	2.523	2.428
25	9.475	6.598	5.462	4.835	4.433	4.150	3.939	3.776	3.645	3.537	3.370	3.196	3.013	2.898	2.819	2.652	2.564	2.519	2.473	2.377
30	9.180	6.355	5.239	4.623	4.228	3.949	3.742	3.580	3.450	3.344	3.179	3.006	2.823	2.708	2.628	2.459	2.370	2.323	2.276	2.176
35	8.976	6.188	5.086	4.479	4.088	3.812	3.607	3.447	3.318	3.212	3.048	2.876	2.693	2.577	2.497	2.327	2.235	2.188	2.139	2.036
40	8.828	6.066	4.976	4.374	3.986	3.713	3.509	3.350	3.222	3.117	2.953	2.781	2.598	2.482	2.401	2.230	2.137	2.088	2.038	1.932
50	8.626	5.902	4.826	4.232	3.849	3.579	3.376	3.219	3.092	2.988	2.825	2.653	2.470	2.353	2.272	2.097	2.001	1.951	1.899	1.786
75	8.366	5.691	4.635	4.050	3.674	3.407	3.208	3.052	2.927	2.823	2.661	2.490	2.306	2.188	2.105	1.925	1.824	1.771	1.714	1.589
100	8.241	5.589	4.542	3.963	3.589	3.325	3.127	2.972	2.847	2.744	2.583	2.411	2.227	2.108	2.024	1.840	1.737	1.681	1.621	1.485
150	8.118	5.490	4.453	3.878	3.508	3.245	3.048	2.894	2.770	2.667	2.506	2.335	2.150	2.030	1.944	1.756	1.649	1.590	1.526	1.374
∞	7.879	5.298	4.279	3.715	3.350	3.091	2.897	2.744	2.621	2.519	2.358	2.187	2.000	1.877	1.789	1.590	1.470	1.402	1.322	(1.0)

Percentage points of the F distribution

$q = 0.999$ | $\alpha_1^R = 0.1\%$ | $\alpha_2 = 0.2\%$ | $\gamma = 99.8\%$

The values for $\nu_2 = 1$ should be multiplied by 10

ν_2 \ ν_1	1	2	3	4	5	6	7	8	9	10	12	15	20	25	30	50	75	100	150	∞
1	40528	50000	54038	56250	57640	58594	59287	59814	60228	60562	61067	61576	62091	62402	62610	63029	63239	63344	63450	63662
2	998.5	999.0	999.2	999.3	999.3	999.3	999.4	999.4	999.4	999.4	999.4	999.4	999.4	999.5	999.5	999.5	999.5	999.5	999.5	999.5
3	167.0	148.5	141.1	137.1	134.6	132.8	131.6	130.6	129.9	129.2	128.3	127.4	126.4	125.8	125.4	124.7	124.3	124.1	123.9	123.5
4	74.14	61.25	56.18	53.44	51.71	50.53	49.66	49.00	48.47	48.05	47.41	46.76	46.10	45.70	45.43	44.88	44.61	44.47	44.33	44.05
5	47.18	37.12	33.20	31.09	29.75	28.83	28.16	27.65	27.24	26.92	26.42	25.91	25.39	25.08	24.87	24.44	24.22	24.12	24.01	23.79
6	35.51	27.00	23.70	21.92	20.80	20.03	19.46	19.03	18.69	18.41	17.99	17.56	17.12	16.85	16.67	16.31	16.12	16.03	15.93	15.75
7	29.25	21.69	18.77	17.20	16.21	15.52	15.02	14.63	14.33	14.08	13.71	13.32	12.93	12.69	12.53	12.20	12.04	11.95	11.87	11.70
8	25.41	18.49	15.83	14.39	13.48	12.86	12.40	12.05	11.77	11.54	11.19	10.84	10.48	10.26	10.11	9.804	9.650	9.571	9.493	9.334
9	22.86	16.39	13.90	12.56	11.71	11.13	10.70	10.37	10.11	9.894	9.570	9.238	8.898	8.689	8.548	8.260	8.113	8.039	7.964	7.813
10	21.04	14.91	12.55	11.28	10.48	9.926	9.517	9.204	8.956	8.754	8.445	8.129	7.804	7.604	7.469	7.193	7.052	6.980	6.908	6.762
11	19.69	13.81	11.56	10.35	9.578	9.047	8.655	8.355	8.116	7.922	7.626	7.321	7.008	6.815	6.684	6.416	6.280	6.210	6.140	5.998
12	18.64	12.97	10.80	9.633	8.892	8.379	8.001	7.710	7.480	7.292	7.005	6.709	6.405	6.217	6.090	5.829	5.695	5.627	5.559	5.420
13	17.82	12.31	10.21	9.073	8.354	7.856	7.489	7.206	6.982	6.799	6.519	6.231	5.934	5.751	5.626	5.370	5.239	5.172	5.104	4.967
14	17.14	11.78	9.729	8.622	7.922	7.436	7.077	6.802	6.583	6.404	6.130	5.848	5.557	5.377	5.254	5.002	4.873	4.807	4.740	4.604
15	16.59	11.34	9.335	8.253	7.567	7.092	6.741	6.471	6.256	6.081	5.812	5.535	5.248	5.071	4.950	4.702	4.573	4.508	4.442	4.307
16	16.12	10.97	9.006	7.944	7.272	6.805	6.460	6.195	5.984	5.812	5.547	5.274	4.992	4.817	4.697	4.451	4.324	4.259	4.193	4.059
17	15.72	10.66	8.727	7.683	7.022	6.562	6.223	5.962	5.754	5.584	5.324	5.054	4.775	4.602	4.484	4.239	4.113	4.049	3.983	3.850
18	15.38	10.39	8.487	7.459	6.808	6.355	6.021	5.763	5.558	5.390	5.132	4.866	4.590	4.418	4.301	4.058	3.933	3.868	3.803	3.670
19	15.08	10.16	8.280	7.265	6.622	6.175	5.845	5.590	5.388	5.222	4.967	4.704	4.430	4.259	4.143	3.902	3.777	3.713	3.647	3.514
20	14.82	9.953	8.098	7.096	6.461	6.019	5.692	5.440	5.239	5.075	4.823	4.562	4.290	4.121	4.005	3.765	3.640	3.576	3.511	3.378
21	14.59	9.772	7.938	6.947	6.318	5.881	5.557	5.308	5.109	4.946	4.696	4.437	4.167	3.999	3.884	3.645	3.520	3.456	3.391	3.257
22	14.38	9.612	7.796	6.814	6.191	5.758	5.438	5.190	4.993	4.832	4.583	4.326	4.058	3.891	3.776	3.538	3.413	3.349	3.284	3.151
23	14.20	9.469	7.669	6.696	6.078	5.649	5.331	5.085	4.890	4.730	4.483	4.227	3.961	3.794	3.680	3.442	3.318	3.254	3.189	3.055
24	14.03	9.339	7.554	6.589	5.977	5.550	5.235	4.991	4.797	4.638	4.393	4.139	3.873	3.707	3.593	3.356	3.232	3.168	3.103	2.969
25	13.88	9.223	7.451	6.493	5.885	5.462	5.148	4.906	4.713	4.555	4.312	4.059	3.794	3.629	3.515	3.279	3.154	3.091	3.025	2.890
30	13.29	8.773	7.054	6.125	5.534	5.122	4.817	4.581	4.393	4.239	4.001	3.753	3.493	3.330	3.217	2.981	2.857	2.792	2.726	2.589
35	12.90	8.470	6.787	5.876	5.298	4.894	4.595	4.363	4.178	4.027	3.792	3.547	3.290	3.128	3.016	2.781	2.655	2.590	2.523	2.383
40	12.61	8.251	6.595	5.698	5.128	4.731	4.436	4.207	4.024	3.874	3.642	3.400	3.145	2.984	2.872	2.636	2.510	2.444	2.376	2.233
50	12.22	7.956	6.336	5.459	4.901	4.512	4.222	3.998	3.818	3.671	3.441	3.204	2.951	2.790	2.670	2.411	2.313	2.246	2.178	2.029
75	11.73	7.585	6.011	5.159	4.617	4.237	3.955	3.736	3.561	3.416	3.192	2.957	2.707	2.547	2.435	2.194	2.062	1.992	1.917	1.754
100	11.50	7.408	5.857	5.017	4.482	4.107	3.829	3.612	3.439	3.296	3.074	2.840	2.591	2.431	2.319	2.076	1.940	1.867	1.790	1.615
150	11.27	7.230	5.707	4.870	4.351	3.081	3.706	3.493	3.321	3.179	2.959	2.727	2.479	2.319	2.206	1.959	1.820	1.744	1.662	1.469
∞	10.83	6.908	5.422	4.617	4.103	3.743	3.475	3.266	3.097	2.959	2.742	2.513	2.266	2.105	1.990	1.733	1.581	1.494	1.395	(1.0)

$q = 0.9999$ | $\alpha_1^R = 0.01\%$ | $\alpha_2 = 0.02\%$ | $\gamma = 99.98\%$

The values for $\nu_2 = 1$ should be multiplied by 1000

ν_2 \ ν_1	1	2	3	4	5	6	7	8	9	10	12	15	20	25	30	50	75	100	150	∞
1	40528	50000	54038	56250	57640	58594	59287	59814	60228	60562	61067	61576	62091	62402	62610	63029	63239	63344	63450	63662
2	9999	9999	9999	9999	9999	9999	9999	9999	9999	9999	9999	9999	9999	9999	9999	9999	9999	9999	9999	9999
3	784.0	694.7	659.3	640.2	628.2	619.9	613.9	609.3	605.7	602.8	598.3	593.8	589.3	586.5	584.7	581.0	579.1	578.1	577.2	575.3
4	241.6	198.0	181.0	171.9	166.1	162.2	159.3	157.1	155.4	154.0	151.9	149.7	147.5	146.2	145.3	143.5	142.6	142.1	141.7	140.8
5	124.9	97.03	86.29	80.53	76.91	74.43	72.61	71.23	70.13	69.25	67.91	66.54	65.16	64.31	63.75	62.60	62.02	61.73	61.43	60.84
6	82.49	61.63	53.68	49.42	46.75	44.91	43.57	42.54	41.73	41.08	40.08	39.07	38.04	37.41	36.98	36.13	35.69	35.47	35.25	34.81
7	62.17	45.13	38.68	35.22	33.06	31.57	30.48	29.64	28.99	28.45	27.64	26.82	25.98	25.46	25.12	24.42	24.06	23.88	23.70	23.34
8	50.69	36.00	30.46	27.49	25.63	24.36	23.42	22.71	22.14	21.68	20.98	20.27	19.55	19.10	18.80	18.19	17.89	17.73	17.57	17.26
9	43.48	30.34	25.40	22.77	21.11	19.97	19.14	18.50	18.00	17.59	16.97	16.33	15.68	15.28	15.01	14.47	14.19	14.05	13.91	13.62
10	38.58	26.55	22.04	19.63	18.12	17.08	16.32	15.74	15.27	14.90	14.33	13.75	13.15	12.78	12.54	12.03	11.77	11.65	11.51	11.25
11	35.06	23.85	19.66	17.42	16.02	15.05	14.34	13.80	13.37	13.02	12.49	11.95	11.39	11.05	10.81	10.34	10.10	9.977	9.854	9.605
12	32.43	21.85	17.90	15.79	14.47	13.56	12.89	12.38	11.98	11.65	11.14	10.63	10.10	9.777	9.557	9.108	8.878	8.762	8.644	8.406
13	30.39	20.31	16.55	14.55	13.29	12.42	11.79	11.30	10.92	10.60	10.12	9.632	9.127	8.816	8.606	8.175	7.954	7.842	7.729	7.500
14	28.77	19.09	15.49	13.57	12.37	11.53	10.92	10.46	10.09	9.785	9.325	8.853	8.366	8.067	7.864	7.448	7.234	7.126	7.016	6.793
15	27.45	18.11	14.64	12.78	11.62	10.82	10.23	9.780	9.422	9.131	8.686	8.229	7.758	7.468	7.271	6.866	6.658	6.553	6.446	6.229
16	26.36	17.30	13.93	12.14	11.01	10.23	9.663	9.226	8.878	8.596	8.164	7.720	7.262	6.979	6.787	6.392	6.189	6.086	5.981	5.768
17	25.44	16.62	13.34	11.60	10.50	9.747	9.191	8.765	8.427	8.152	7.730	7.297	6.850	6.573	6.385	5.999	5.799	5.698	5.595	5.385
18	24.66	16.04	12.85	11.14	10.07	9.335	8.792	8.376	8.046	7.777	7.365	6.941	6.503	6.232	6.047	5.667	5.471	5.371	5.270	5.063
19	23.99	15.55	12.42	10.75	9.706	8.983	8.452	8.044	7.720	7.457	7.053	6.637	6.207	5.941	5.759	5.385	5.191	5.093	4.993	4.788
20	23.40	15.12	12.05	10.41	9.388	8.679	8.158	7.757	7.439	7.181	6.704	6.375	5.952	5.680	5.610	5.141	4.950	4.852	4.753	4.550
21	22.89	14.74	11.73	10.12	9.111	8.414	7.901	7.507	7.195	6.940	6.549	6.147	5.729	5.471	5.294	4.929	4.740	4.643	4.545	4.344
22	22.43	14.41	11.44	9.860	8.867	8.180	7.676	7.288	6.980	6.729	6.343	5.946	5.534	5.279	5.104	4.743	4.555	4.459	4.362	4.162
23	22.03	14.12	11.19	9.630	8.651	7.974	7.476	7.093	6.789	6.542	6.161	5.769	5.362	5.109	4.936	4.578	4.392	4.297	4.200	4.000
24	21.66	13.85	10.96	9.425	8.458	7.790	7.298	6.920	6.620	6.375	5.999	5.611	5.208	4.958	4.787	4.432	4.247	4.152	4.055	3.857
25	21.34	13.62	10.76	9.240	8.285	7.624	7.138	6.765	6.468	6.226	5.854	5.470	5.071	4.823	4.653	4.300	4.116	4.022	3.926	3.728
30	20.09	12.72	9.994	8.544	7.632	7.002	6.537	6.180	5.896	5.664	5.308	4.939	4.554	4.314	4.149	3.806	3.625	3.532	3.437	3.240
35	19.26	12.12	9.487	8.084	7.202	6.592	6.143	5.796	5.521	5.296	4.950	4.591	4.216	3.981	3.819	3.481	3.303	3.210	3.115	2.918
40	18.67	11.70	9.128	7.759	6.899	6.303	5.864	5.526	5.256	5.036	4.697	4.345	3.977	3.746	3.587	3.252	3.074	2.982	2.007	2.600
50	17.88	11.14	8.652	7.330	6.498	5.922	5.497	5.170	4.909	4.695	4.366	4.024	3.664	3.438	3.281	2.950	2.773	2.680	2.584	2.380
75	16.89	10.44	8.066	6.802	6.006	5.455	5.048	4.734	4.483	4.278	3.961	3.630	3.281	3.060	2.907	2.578	2.399	2.304	2.205	1.988
100	16.43	10.11	7.791	6.555	5.777	5.237	4.839	4.531	4.285	4.084	3.773	3.448	3.104	2.885	2.732	2.404	2.223	2.126	2.024	1.795
150	15.98	9.800	7.528	6.319	5.558	5.030	4.640	4.338	4.097	3.900	3.594	3.274	2.934	2.718	2.566	2.236	2.052	1.953	1.846	1.597
∞	15.14	9.210	7.036	5.878	5.149	4.643	4.268	3.978	3.747	3.556	3.261	2.951	2.619	2.406	2.254	1.919	1.724	1.613	1.487	(1.0)

Critical values for the Kolmogorov–Smirnov goodness-of-fit test (for completely specified distributions)

α_1	5%	2½%	1%	½%
α_2	10%	5%	2%	1%
n				
1	0.9500	0.9750	0.9900	0.9950
2	0.7764	0.8419	0.9000	0.9293
3	0.6360	0.7076	0.7846	0.8290
4	0.5652	0.6239	0.6889	0.7342
5	0.5094	0.5633	0.6272	0.6685
6	0.4680	0.5193	0.5774	0.6166
7	0.4361	0.4834	0.5384	0.5758
8	0.4096	0.4543	0.5065	0.5418
9	0.3875	0.4300	0.4796	0.5133
10	0.3687	0.4092	0.4566	0.4889
11	0.3524	0.3912	0.4367	0.4677
12	0.3382	0.3754	0.4192	0.4490
13	0.3255	0.3614	0.4036	0.4325
14	0.3142	0.3489	0.3897	0.4176
15	0.3040	0.3376	0.3771	0.4042
16	0.2947	0.3273	0.3657	0.3920
17	0.2863	0.3180	0.3553	0.3809
18	0.2785	0.3094	0.3457	0.3706
19	0.2714	0.3014	0.3369	0.3612
20	0.2647	0.2941	0.3287	0.3524

α_1	5%	2½%	1%	½%
α_2	10%	5%	2%	1%
n				
21	0.2586	0.2872	0.3210	0.3443
22	0.2528	0.2809	0.3139	0.3367
23	0.2475	0.2749	0.3073	0.3295
24	0.2424	0.2693	0.3010	0.3229
25	0.2377	0.2640	0.2952	0.3166
26	0.2332	0.2591	0.2896	0.3106
27	0.2290	0.2544	0.2844	0.3050
28	0.2250	0.2499	0.2794	0.2997
29	0.2212	0.2457	0.2747	0.2947
30	0.2176	0.2417	0.2702	0.2899
31	0.2141	0.2379	0.2660	0.2853
32	0.2108	0.2342	0.2619	0.2809
33	0.2077	0.2308	0.2580	0.2768
34	0.2047	0.2274	0.2543	0.2728
35	0.2018	0.2242	0.2507	0.2690
36	0.1991	0.2212	0.2473	0.2653
37	0.1965	0.2183	0.2440	0.2618
38	0.1939	0.2154	0.2409	0.2584
39	0.1915	0.2127	0.2379	0.2552
40	0.1891	0.2101	0.2349	0.2521

α_1	5%	2½%	1%	½%
α_2	10%	5%	2%	1%
n				
41	0.1869	0.2076	0.2321	0.2490
42	0.1847	0.2052	0.2294	0.2461
43	0.1826	0.2028	0.2268	0.2433
44	0.1805	0.2006	0.2243	0.2406
45	0.1786	0.1984	0.2218	0.2380
46	0.1767	0.1963	0.2194	0.2354
47	0.1748	0.1942	0.2171	0.2330
48	0.1730	0.1922	0.2149	0.2306
49	0.1713	0.1903	0.2128	0.2283
50	0.1696	0.1884	0.2107	0.2260
55	0.1619	0.1798	0.2011	0.2157
60	0.1551	0.1723	0.1927	0.2067
65	0.1491	0.1657	0.1853	0.1988
70	0.1438	0.1597	0.1786	0.1917
75	0.1390	0.1544	0.1727	0.1853
80	0.1347	0.1496	0.1673	0.1795
85	0.1307	0.1452	0.1624	0.1742
90	0.1271	0.1412	0.1579	0.1694
95	0.1238	0.1375	0.1537	0.1649
100	0.1207	0.1340	0.1499	0.1608

Goodness-of-fit tests are designed to test a null hypothesis that some given data are a random sample from a specified probability distribution. The Kolmogorov–Smirnov tests are based on the maximum absolute difference D_n between the c.d.f. (cumulative distribution function) $F_0(x)$ of the hypothesised distribution and the c.d.f. of the sample (sometimes called the empirical c.d.f.) $F_n(x)$. This sample c.d.f. is the step-function which starts at 0 and rises by $1/n$ at each observed value, where n is the sample size; i.e. $F_n(x)$ is equal to the proportion of the sample values which are less than or equal to x.

Critical regions for rejecting H_0 are of the form $D_n \geqslant$ *tabulated value*, and in most cases the general alternative hypothesis is appropriate, i.e. the α_2 significance levels should be used. One-sided alternative hypotheses can be dealt with by only considering differences *in one direction* between the c.d.f.s. For example, suppose H_1 says that the actual values being sampled are mainly *less* than those expected from $F_0(x)$. If this is the case $F_n(x)$ will tend to rise earlier than $F_0(x)$, and so instead of D_n we should then use the statistic $D_n^+ = \max\{F_n(x) - F_0(x)\}$. In the opposite case, where H_1 says that the values sampled are mainly *greater* than those expected from $F_0(x)$, we should use $D_n^- = \max\{F_0(x) - F_n(x)\}$. Critical regions are D_n^+ (or D_n^-) $\geqslant$ *tabulated value*, and in these one-sided tests the α_1 significance levels should be used.

For illustration, let us test the null hypothesis H_0 that the following ten observations (derived in fact from part of the top row of the table of random digits on page 42) are a random sample from the uniform distribution over $(0:1)$, having c.d.f. $F_0(x) = 0$ for $x < 0$, $F_0(x) = x$ for $0 \leqslant x \leqslant 1$, and $F_0(x) = 1$ for $x > 1$:

0.02484	0.88139	0.31788	0.35873	0.63259	0.99886	0.20644	0.41853	0.41915	0.02944

Sorting the data into ascending order, we have:

0.02484	0.02944	0.20644	0.31788	0.35873	0.41853	0.41915	0.63259	0.88139	0.99886

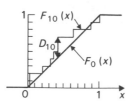

It is then easy to draw the sample c.d.f., $F_{10}(x)$, and from the diagram we find that the maximum vertical distance between the two c.d.f.s, which occurs at $x = 0.41915$, is $D_{10} = 0.7 - 0.41915 = 0.28085$. But the critical region for rejection of H_0 even at the $\alpha_2 = 10\%$ significance level is $D_{10} \geqslant 0.3687$, and so we have no reason here to doubt the null hypothesis.

The Kolmogorov–Smirnov test may be used both when $F_0(x)$ is continuous and discrete. In the continuous case critical values are exact; in the discrete case they may be conservative (i.e. true $\alpha <$ nominal α).

A particularly useful application of the test is to test data for normality. In this case use may be made of the graph on page 27 of the c.d.f. of the standard normal distribution by first standardising the data, i.e. subtracting the mean and dividing by the standard deviation. The resulting sample c.d.f. may be drawn on page 27 and the Kolmogorov–Smirnov test performed as usual. For example to test the hypothesis that the following data come from the normal distribution with mean 5 and standard deviation 2, we transform each observation X into $Z = \frac{1}{2}(X - 5)$:

(original) X	8.74	4.08	8.31	7.80	6.39	7.21	7.05	5.94
(transformed) Z	1.87	−0.46	1.655	1.40	0.695	1.105	1.025	0.47

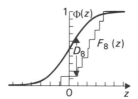

Then we sort the transformed data into ascending order and draw the sample c.d.f. on the graph on page 27 (step-heights are $1/8$ since the sample size n is 8 here). The maximum vertical distance between the two c.d.f.s is seen to be about 0.556, and this shows strong evidence that the data do not come from the hypothesised distribution, since the $\alpha_2 = 1\%$ critical region is $D_8 \geqslant 0.5418$.

Perhaps it is more commonly necessary to test for normality *without* the mean and standard deviation being specified. To perform the test in these circumstances, first estimate the mean by $\bar{X} = \Sigma X/n$ and the standard deviation by $s = \{\Sigma(X - \bar{X})^2/(n-1)\}^{1/2}$. Standardise the data using these estimates, and then proceed as before except that the critical values on page 27 should be used. For the above eight observations, $\bar{X} = 6.940$ and $s = 1.484$. The transformed data are now:

1.213	−1.927	0.923	0.579	−0.371	0.182	0.074	−0.674

The maximum difference now found between the c.d.f. of this sample and that of the standard normal distribution is $D_8 = 0.155$, and this is certainly not significantly large, for even at the $\alpha_2 = 10\%$ level the critical region is $D_8 \geqslant 0.2652$. We conclude therefore that although there was strong evidence that the data do not come from the originally specified normal distribution, they could quite easily have come from some other normal distribution. The originator of this type of test was W. H. Lilliefors.

Critical values for larger sample sizes than covered in the tables are discussed on page 35.

Critical values for the Kolmogorov–Smirnov test for normality

α_1	5%	2½%	1%	½%
α_2	10%	5%	2%	1%
n				
1	–	–	–	–
2	–	–	–	–
3	0.3666	0.3758	0.3812	0.3830
4	0.3453	0.3753	0.4007	0.4131
5	0.3189	0.3431	0.3755	0.3970
6	0.2972	0.3234	0.3523	0.3708
7	0.2802	0.3043	0.3321	0.3509
8	0.2652	0.2880	0.3150	0.3332
9	0.2523	0.2741	0.2999	0.3174
10	0.2411	0.2619	0.2869	0.3037
11	0.2312	0.2514	0.2754	0.2916
12	0.2225	0.2420	0.2651	0.2810
13	0.2148	0.2336	0.2559	0.2714
14	0.2077	0.2261	0.2476	0.2627
15	0.2013	0.2192	0.2401	0.2549
16	0.1954	0.2129	0.2332	0.2476
17	0.1901	0.2071	0.2270	0.2410
18	0.1852	0.2017	0.2212	0.2349
19	0.1807	0.1968	0.2158	0.2292
20	0.1765	0.1921	0.2107	0.2238

α_1	5%	2½%	1%	½%
α_2	10%	5%	2%	1%
n				
21	0.1725	0.1878	0.2060	0.2188
22	0.1688	0.1838	0.2015	0.2141
23	0.1653	0.1800	0.1974	0.2097
24	0.1620	0.1764	0.1936	0.2056
25	0.1589	0.1730	0.1899	0.2018
26	0.1560	0.1699	0.1865	0.1981
27	0.1533	0.1670	0.1833	0.1947
28	0.1507	0.1642	0.1802	0.1915
29	0.1483	0.1615	0.1773	0.1884
30	0.1460	0.1589	0.1746	0.1855
31	0.1437	0.1565	0.1719	0.1827
32	0.1416	0.1542	0.1693	0.1800
33	0.1395	0.1519	0.1669	0.1774
34	0.1375	0.1498	0.1645	0.1749
35	0.1356	0.1478	0.1622	0.1725
36	0.1338	0.1458	0.1601	0.1702
37	0.1321	0.1439	0.1580	0.1680
38	0.1304	0.1421	0.1560	0.1659
39	0.1288	0.1403	0.1540	0.1638
40	0.1272	0.1386	0.1522	0.1618

α_1	5%	2½%	1%	½%
α_2	10%	5%	2%	1%
n				
41	0.1257	0.1370	0.1504	0.1599
42	0.1243	0.1354	0.1487	0.1581
43	0.1229	0.1339	0.1470	0.1563
44	0.1216	0.1325	0.1454	0.1546
45	0.1203	0.1311	0.1438	0.1530
46	0.1190	0.1297	0.1423	0.1514
47	0.1178	0.1284	0.1409	0.1498
48	0.1166	0.1271	0.1394	0.1483
49	0.1155	0.1258	0.1380	0.1468
50	0.1144	0.1246	0.1367	0.1454
55	0.1092	0.1190	0.1306	0.1389
60	0.1048	0.1142	0.1253	0.1332
65	0.1008	0.1098	0.1205	0.1281
70	0.0972	0.1060	0.1163	0.1236
75	0.0940	0.1025	0.1125	0.1195
80	0.0911	0.0993	0.1090	0.1158
85	0.0885	0.0964	0.1059	0.1125
90	0.0861	0.0938	0.1030	0.1094
95	0.0838	0.0913	0.1003	0.1065
100	0.0817	0.0890	0.0978	0.1039

For description see page 26; for larger sample sizes, see page 35.

The c.d.f. of the standard normal distribution

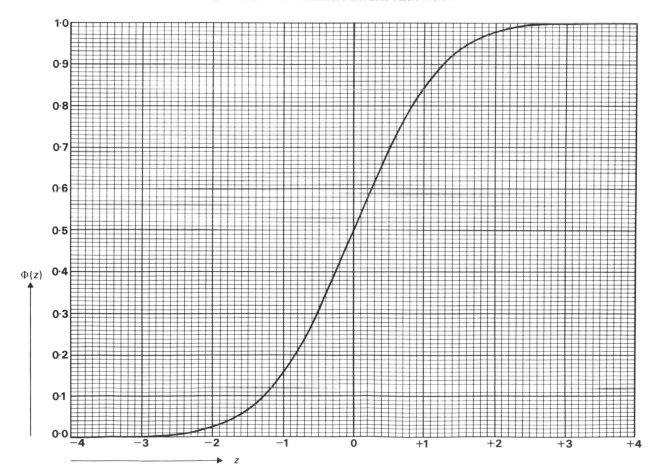

Nonparametric tests

Pages 29–34 give critical values for six nonparametric tests. The sign test and the Wilcoxon signed-rank test are one-sample tests and can also be applied to matched-pairs data, the Mann–Whitney and Kolmogorov–Smirnov tests are two-sample tests, and the Kruskal–Wallis and Friedman tests are nonparametric alternatives to the standard one-way and two-way analyses-of-variance. Critical values for larger sample sizes than those included in these tables are covered on page 35.

The sign test (page 29). Suppose that the national average mark in an English examination is 60%. (In nonparametric work, the *average* is usually taken to be the median rather than the mean.) Test whether the following marks, obtained by twelve students from a particular school, are consistent with this average.

70	65	75	58	56	60	80	75	71	69	58	75
+	+	+	−	−	0	+	+	+	+	−	+

We have printed + or − under each mark to indicate whether it is greater or less than the hypothesised 60. There is one mark of exactly 60 which is completely ignored for the purposes of the test, reducing the sample size n to 11. The sign test statistic S is the number of + signs or the number of − signs, whichever is smaller; here $S = 3$. Critical regions are of the form $S \leqslant$ *tabulated value*. As the $\alpha_2 = 10\%$ critical region for $n = 11$ is $S \leqslant 2$, we cannot reject the null hypothesis H_0 that these marks are consistent with an average of 60%.

For a one-sided test, count either the number of + or − signs, *whichever the alternative hypothesis H_1 suggests should be the smaller*. For example if H_1 says that the average mark is *less* than 60%, S would be defined as the number of + signs since if H_1 is true there will generally be fewer marks exceeding 60%. Critical regions are of the same form as previously, but the α_1 significance levels should be used.

The Wilcoxon signed-rank test (page 29). This test is more powerful than the sign test as it takes into account the sizes of the differences from the hypothesised average, rather than just their signs. In the above example, first subtract 60 from each mark, and then rank the resulting differences, irrespective of their signs. Again ignore the mark of exactly 60, and also average the ranks of tied observations.

differences	+10	+5	+15	−2	−4	(0)	+20	+15	+11	+9	−2	+15
ranks	6	4	9	$1\frac{1}{2}$	3		11	9	7	5	$1\frac{1}{2}$	9

The Wilcoxon statistic T is the sum of the ranks of either the +ve or −ve differences, whichever is smaller. Here $T = 1\frac{1}{2} + 3 + 1\frac{1}{2} = 6$. Critical regions are of the form $T \leqslant$ *tabulated value*, and the test thus shows evidence at better than the $\alpha_2 = 2\%$ significance level that these marks are inconsistent with the national average, since the 2% critical region for $n = 11$ is $T \leqslant 7$.

For a one-sided test, let T be the sum of the ranks of either the +ve or the −ve differences, whichever the one-sided H_1 suggests should be the smaller — it will be the same choice as in the sign test — and use the α_1 significance levels.

Matched-pairs data. Matched-pairs data arise in such examples as the following. One member of each of eight pairs of identical twins is taught mathematics by programmed learning, the other by a standard teaching method. Do the test results imply any difference in the effectiveness of the two teaching methods?

twins	a	b	c	d	e	f	g	h
programmed learning	70	80	62	50	70	30	49	60
standard method	75	82	65	58	68	41	55	67
differences	+5	+2	+3	+8	−2	+11	+6	+7

Such data may be analysed by either of the above tests, comparing the twin-by-twin differences in the final row with a hypothesised average of 0. The reader may confirm that $S = 1$ and $T = 1\frac{1}{2}$, so that the null hypothesis of no difference is rejected at the $\alpha_2 = 10\%$ level in the sign test and at near to the $\alpha_2 = 2\%$ level in Wilcoxon's test.

The Mann–Whitney U test (page 30). Six students from another school take the same English examination as mentioned above. Their marks are: 53, 65, 63, 57, 68 and 56. We want to check whether the two sets of students are of different average standards.

We order the two samples of marks together and indicate by A or B whether a mark comes from the first or second school:

53	56	56	57	58	58	60	63	65	65	68	69	70	71	75	75	75
B	A	B	B	A	A	B	B	A	B	B	A	A	A	A	A	A
ranks 1	$2\frac{1}{2}$	$2\frac{1}{2}$	4	5	6	7	8	$9\frac{1}{2}$	$9\frac{1}{2}$	11	12	13	14	15	16	17

The observations are given ranks as shown, the ranks being averaged in the case of ties (unnecessary if a tie only involves members of one sample). Then either form the sum R_A of the ranks of observations from sample A, and calculate $U_A = R_A - \frac{1}{2}n_A(n_A + 1)$, or the sum R_B of the ranks of observations from sample B, and calculate $U_B = R_B - \frac{1}{2}n_B(n_B + 1)$, where n_A and n_B are the sizes of samples A and B. Finally obtain U as the smaller of U_A or $n_A n_B - U_A$, or equivalently the smaller of U_B or $n_A n_B - U_B$. Critical regions have the form $U \leqslant$ *tabulated value*. In the above example, $R_A = 135$ so that $U_A = 135 - \frac{1}{2}(12)(13) = 57$, or $R_B = 36$ and $U_B = 36 - \frac{1}{2}(6)(7) = 15$. In either case U is found to be 15, and this provides a little evidence for a difference between the two sets of students since the $\alpha_2 = 10\%$ critical region is $U \leqslant 17$ and the 5% region is $U \leqslant 14$. (In the table, sample sizes are denoted by n_1 and n_2 with $n_1 \leqslant n_2$.)

For a one-sided test, calculate whichever of U_A and U_B is more likely to be small if the one-sided H_1 is true, use this in place of U, and refer to the α_1 significance levels.

The Kolmogorov–Smirnov two-sample test (page 31). Whereas the Mann–Whitney test is designed specifically to detect differences in average, the Kolmogorov–Smirnov test is used when other types of difference may also be of interest. To calculate the test statistic D, draw the sample c.d.f.s (see page 26) for both sample A and sample B on the same graph; D is then the maximum vertical distance between these two c.d.f.s. To use the table on page 31, form $D^* = n_A n_B D$, and critical regions are of the form $D^* \geqslant$ *tabulated value*, using the α_2 significance levels. A one-sided version of the test is also available, but is not often used since the alternative hypothesis is then essentially concerned not with general differences but a difference in average, for which the Mann–Whitney test is more powerful. Applied to the above example on the two sets of English results, $D = 7/12$ and $D^* = 12 \times 6 \times 7/12 = 42$. This is not even significant at the $\alpha_2 = 10\%$ level, as that critical region is $D^* \geqslant 48$. This supports the above remark that the Mann–Whitney test (which gave significance at better than the 10% level) is more powerful as a test for differences in average.

The Kruskal–Wallis test (pages 32–34). The Kruskal–Wallis test is also designed to detect differences in average, but now when we have three or more samples to compare. Again, as in the Mann–Whitney test, we rank all of the data together (averaging the ranks of tied observations) and form the sum of the ranks in each sample. The test statistic is

$$H = \frac{12}{N(N + 1)} \sum_{i=1}^{k} \frac{R_i^2}{n_i} - 3(N + 1)$$

where k is the number of samples, $n_1, n_2, \ldots, n_k$ are their sizes, $N = \Sigma n_i$ and $R_1, R_2, \ldots, R_k$ are the rank sums. Critical regions are of the form $H \geqslant$ *tabulated value*. Tables are given on page 32 for $k = 3$ and $N \leqslant 19$, on page 33 for $k = 4$ ($N \leqslant 14$), $k = 5$ ($N \leqslant 13$) and $k = 6$ ($N \leqslant 13$), and on page 34 for $3 \leqslant k \leqslant 6$ and equal sample sizes $n_1 = n_2 = \ldots = n_k = n$ for $2 \leqslant n \leqslant 25$.

To illustrate the Kruskal–Wallis test, we show samples of mileages per gallon for three different engine designs:

design	mileage per gallon				ranks				rank sums
a	19.8	20.5	20.8	19.7	4	6	$7\frac{1}{2}$	$2\frac{1}{2}$	20
b	21.7	20.8	21.2		10	$7\frac{1}{2}$	9		$26\frac{1}{2}$
c	19.7	19.4	19.9		$2\frac{1}{2}$	1	5		$8\frac{1}{2}$

Then

$$H = \frac{12}{10 \times 11} \left(\frac{20^2}{4} + \frac{(26\frac{1}{2})^2}{3} + \frac{(8\frac{1}{2})^2}{3} \right) - 3 \times 11$$

$$= 0.1091 \times (358.167) - 33 = 6.073.$$

This is significant of a difference between average mileages at better than the 5% level, the $\alpha = 5\%$ critical region being $H \geqslant 5.791$. (In such cases where there is no meaningful one-sided version of the test, α_2 is written as α with no subscript.)

Friedman's test (page 34). Friedman's test applies when the observations in three or more samples are related or 'blocked' (similarly as with matched-pairs data). If there are k samples and n blocks, the observations in each block are ranked from 1 to k, the rank sums $R_1, R_2, \ldots, R_k$ for each sample obtained, and Friedman's test statistic is then

$$M = \frac{12}{nk(k + 1)} \sum_{i=1}^{k} R_i^2 - 3n(k + 1)$$

To illustrate the test, suppose that in a mileages survey we use cars of five different ages and obtain the following data:

sign	age of car 1	2	3	4	5	ranks					rank sums
a	21.3	21.6	21.2	20.7	20.1	2	2	2½	2	2	10½
b	21.6	21.7	21.2	20.9	20.6	3	3	2½	3	3	14½
c	20.0	20.1	19.9	19.5	19.0	1	1	1	1	1	5

Then $M = 12/(15 \times 4)\{(10\tfrac{1}{2})^2 + (14\tfrac{1}{2})^2 + 5^2\}$ $(15 \times 4) = 0.2 \times 345.5 - 60 = 9.1$, which is strongly significant since the $\alpha = 1\%$ critical region is $M \geq 8.400$.

Note: All of the nonparametric tests described above have discrete-valued statistics, so that the exact nominal α-levels are not usually obtainable. The tables give *best conservative* critical regions, i.e. the largest regions with significance levels less than or equal to α.

Critical values for the sign test

α_1	5%	2½%	1%	½%
α_2	10%	5%	2%	1%
n				
1	—	—	—	—
2	—	—	—	—
3	—	—	—	—
4	—	—	—	—
5	0	—	—	—
6	0	0	—	—
7	0	0	0	—
8	1	0	0	0
9	1	1	0	0
10	1	1	0	0
11	2	1	1	0
12	2	2	1	1
13	3	2	1	1
14	3	2	2	1
15	3	3	2	2
16	4	3	2	2
17	4	4	3	2
18	5	4	3	3
19	5	4	4	3
20	5	5	4	3
21	6	5	4	4
22	6	5	5	4
23	7	6	5	4
24	7	6	5	5
25	7	7	6	5
26	8	7	6	6
27	8	7	7	6
28	9	8	7	6
29	9	8	7	7
30	10	9	8	7
31	10	9	8	7
32	10	9	8	8
33	11	10	9	8
34	11	10	9	9
35	12	11	10	9
36	12	11	10	9
37	13	12	10	10
38	13	12	11	10
39	13	12	11	11
40	14	13	12	11
41	14	13	12	11
42	15	14	13	12
43	15	14	13	12
44	16	15	13	13
45	16	15	14	13
46	16	15	14	13
47	17	16	15	14
48	17	16	15	14
49	18	17	15	15
50	18	17	16	15
51	19	18	16	15
52	19	18	17	16
53	20	18	17	16
54	20	19	18	17
55	20	19	18	17
56	21	20	18	17
57	21	20	19	18
58	22	21	19	18
59	22	21	20	19
60	23	21	20	19
61	23	22	20	20
62	24	22	21	20
63	24	23	21	20
64	24	23	22	21
65	25	24	22	21
66	25	24	23	22
67	26	25	23	22
68	26	25	23	22
69	27	25	24	23
70	27	26	24	23
71	28	26	25	24
72	28	27	25	24
73	28	27	26	25
74	29	28	26	25
75	29	28	26	25
76	30	28	27	26
77	30	29	27	26
78	31	29	28	27
79	31	30	28	27
80	32	30	29	28
81	32	31	29	28
82	33	31	30	28
83	33	32	30	29
84	33	32	30	29
85	34	32	31	30
86	34	33	31	30
87	35	33	32	31
88	35	34	32	31
89	36	34	33	31
90	36	35	33	32
91	37	35	33	32
92	37	36	34	33
93	38	36	34	33
94	38	37	35	34
95	38	37	35	34
96	39	37	36	34
97	39	38	36	35
98	40	38	37	35
99	40	39	37	36
100	41	39	37	36

For description, see page 28; for larger sample sizes, see page 35.

Critical values for the Wilcoxon signed-rank test

α_1	5%	2½%	1%	½%
α_2	10%	5%	2%	1%
n				
1	—	—	—	—
2	—	—	—	—
3	—	—	—	—
4	—	—	—	—
5	0	—	—	—
6	2	0	—	—
7	3	2	0	—
8	5	3	1	0
9	8	5	3	1
10	10	8	5	3
11	13	10	7	5
12	17	13	9	7
13	21	17	12	9
14	25	21	15	12
15	30	25	19	15
16	35	29	23	19
17	41	34	27	23
18	47	40	32	27
19	53	46	37	32
20	60	52	43	37
21	67	58	49	42
22	75	65	55	48
23	83	73	62	54
24	91	81	69	61
25	100	89	76	68
26	110	98	84	75
27	119	107	92	83
28	130	116	101	91
29	140	126	110	100
30	151	137	120	109
31	163	147	130	118
32	175	159	140	128
33	187	170	151	138
34	200	182	162	148
35	213	195	173	159
36	227	208	185	171
37	241	221	198	182
38	256	235	211	194
39	271	249	224	207
40	286	264	238	220
41	302	279	252	233
42	319	294	266	247
43	336	310	281	261
44	353	327	296	276
45	371	343	312	291
46	389	361	328	307
47	407	378	345	322
48	426	396	362	339
49	446	415	379	355
50	466	434	397	373
51	486	453	416	390
52	507	473	434	408
53	529	494	454	427
54	550	514	473	445
55	573	536	493	465
56	595	557	514	484
57	618	579	535	504
58	642	602	556	525
59	666	625	578	546
60	690	648	600	567
61	715	672	623	589
62	741	697	646	611
63	767	721	669	634
64	793	747	693	657
65	820	772	718	681
66	847	798	742	705
67	875	825	768	729
68	903	852	793	754
69	931	879	819	779
70	960	907	846	805
71	990	936	873	831
72	1020	964	901	858
73	1050	994	928	884
74	1081	1023	957	912
75	1112	1053	986	940
76	1144	1084	1015	968
77	1176	1115	1044	997
78	1209	1147	1075	1026
79	1242	1179	1105	1056
80	1276	1211	1136	1086
81	1310	1244	1168	1116
82	1345	1277	1200	1147
83	1380	1311	1232	1178
84	1415	1345	1265	1210
85	1451	1380	1298	1242
86	1487	1415	1332	1275
87	1524	1451	1366	1308
88	1561	1487	1400	1342
89	1599	1523	1435	1376
90	1638	1560	1471	1410
91	1676	1597	1507	1445
92	1715	1635	1543	1480
93	1755	1674	1580	1516
94	1795	1712	1617	1552
95	1836	1752	1655	1589
96	1877	1791	1693	1626
97	1918	1832	1731	1664
98	1960	1872	1770	1702
99	2003	1913	1810	1740
100	2045	1955	1850	1779

For description, see page 28; for larger sample sizes, see page 35.

Critical values for the Mann–Whitney U test

α_1 / n_1	n_2	5% / 10%	2½% / 5%	1% / 2%	½% / 1%
2	2	–	–	–	–
2	3	–	–	–	–
2	4	–	–	–	–
2	5	0	–	–	–
2	6	0	–	–	–
2	7	0	–	–	–
2	8	1	0	–	–
2	9	1	0	–	–
2	10	1	0	–	–
2	11	1	0	–	–
2	12	2	1	–	–
2	13	2	1	0	–
2	14	3	1	0	–
2	15	3	1	0	–
2	16	3	1	0	–
2	17	3	2	0	–
2	18	4	2	0	–
2	19	4	2	1	0
2	20	4	2	1	0
2	21	5	3	1	0
2	22	5	3	1	0
2	23	5	3	1	0
2	24	6	3	1	0
2	25	6	3	1	0
3	3	0	–	–	–
3	4	0	–	–	–
3	5	1	0	–	–
3	6	2	1	–	–
3	7	2	1	0	–
3	8	3	2	0	–
3	9	4	2	1	0
3	10	4	3	1	0
3	11	5	3	1	0
3	12	5	4	2	1
3	13	6	4	2	1
3	14	7	5	2	1
3	15	7	5	3	2
3	16	8	6	3	2
3	17	9	6	4	2
3	18	9	7	4	2
3	19	10	7	4	3
3	20	11	8	5	3
3	21	11	8	5	3
3	22	12	9	6	4
3	23	13	9	6	4
3	24	13	10	6	4
3	25	14	10	7	5
4	4	1	0	–	–
4	5	2	1	0	–
4	6	3	2	1	0
4	7	4	3	1	0
4	8	5	4	2	1
4	9	6	4	3	1
4	10	7	5	3	2
4	11	8	6	4	2
4	12	9	7	5	3
4	13	10	8	5	3
4	14	11	9	6	4
4	15	12	10	7	5
4	16	14	11	7	5
4	17	15	11	8	6
4	18	16	12	9	6
4	19	17	13	9	7
4	20	18	14	10	8
4	21	19	15	11	8
4	22	20	16	11	9
4	23	21	17	12	9
4	24	22	17	13	10
4	25	23	18	13	10

n_1	n_2	5% / 10%	2½% / 5%	1% / 2%	½% / 1%
5	5	4	2	1	0
5	6	5	3	2	1
5	7	6	5	3	1
5	8	8	6	4	2
5	9	9	7	5	3
5	10	11	8	6	4
5	11	12	9	7	5
5	12	13	11	8	6
5	13	15	12	9	7
5	14	16	13	10	7
5	15	18	14	11	8
5	16	19	15	12	9
5	17	20	17	13	10
5	18	22	18	14	11
5	19	23	19	15	12
5	20	25	20	16	13
5	21	26	22	17	14
5	22	28	23	18	14
5	23	29	24	19	15
5	24	30	25	20	16
5	25	32	27	21	17
6	6	7	5	3	2
6	7	8	6	4	3
6	8	10	8	6	4
6	9	12	10	7	5
6	10	14	11	8	6
6	11	16	13	9	7
6	12	17	14	11	9
6	13	19	16	12	10
6	14	21	17	13	11
6	15	23	19	15	12
6	16	25	21	16	13
6	17	26	22	18	15
6	18	28	24	19	16
6	19	30	25	20	17
6	20	32	27	22	18
6	21	34	29	23	19
6	22	36	30	24	21
6	23	37	32	26	22
6	24	39	33	27	23
6	25	41	35	29	24
7	7	11	8	6	4
7	8	13	10	7	6
7	9	15	12	9	7
7	10	17	14	11	9
7	11	19	16	12	10
7	12	21	18	14	12
7	13	24	20	16	13
7	14	26	22	17	15
7	15	28	24	19	16
7	16	30	26	21	18
7	17	33	28	23	19
7	18	35	30	24	21
7	19	37	32	26	22
7	20	39	34	28	24
7	21	41	36	30	25
7	22	44	38	31	27
7	23	46	40	33	29
7	24	48	42	35	30
7	25	50	44	36	32
8	8	15	13	9	7
8	9	18	15	11	9
8	10	20	17	13	11
8	11	23	19	15	13
8	12	26	22	17	15
8	13	28	24	20	17
8	14	31	26	22	18
8	15	33	29	24	20

n_1	n_2	5% / 10%	2½% / 5%	1% / 2%	½% / 1%
8	16	36	31	26	22
8	17	39	34	28	24
8	18	41	36	30	26
8	19	44	38	32	28
8	20	47	41	34	30
8	21	49	43	36	32
8	22	52	45	38	34
8	23	54	48	40	35
8	24	57	50	42	37
8	25	60	53	45	39
9	9	21	17	14	11
9	10	24	20	16	13
9	11	27	23	18	16
9	12	30	26	21	18
9	13	33	28	23	20
9	14	36	31	26	22
9	15	39	34	28	24
9	16	42	37	31	27
9	17	45	39	33	29
9	18	48	42	36	31
9	19	51	45	38	33
9	20	54	48	40	36
9	21	57	50	43	38
9	22	60	53	45	40
9	23	63	56	48	43
9	24	66	59	50	45
9	25	69	62	53	47
10	10	27	23	19	16
10	11	31	26	22	18
10	12	34	29	24	21
10	13	37	33	27	24
10	14	41	36	30	26
10	15	44	39	33	29
10	16	48	42	36	31
10	17	51	45	38	34
10	18	55	48	41	37
10	19	58	52	44	39
10	20	62	55	47	42
10	21	65	58	50	44
10	22	68	61	53	47
10	23	72	64	55	50
10	24	75	67	58	52
10	25	79	71	61	55
11	11	34	30	25	21
11	12	38	33	28	24
11	13	42	37	31	27
11	14	46	40	34	30
11	15	50	44	37	33
11	16	54	47	41	36
11	17	57	51	44	39
11	18	61	55	47	42
11	19	65	58	50	45
11	20	69	62	53	48
11	21	73	65	57	51
11	22	77	69	60	54
11	23	81	73	63	57
11	24	85	76	66	60
11	25	89	80	70	63
12	12	42	37	31	27
12	13	47	41	35	31
12	14	51	45	38	34
12	15	55	49	42	37
12	16	60	53	46	41
12	17	64	57	49	44
12	18	68	61	53	47
12	19	72	65	56	51
12	20	77	69	60	54

n_1	n_2	5% / 10%	2½% / 5%	1% / 2%	½% / 1%
12	21	81	73	64	58
12	22	85	77	67	61
12	23	90	81	71	64
12	24	94	85	75	68
12	25	98	89	78	71
13	13	51	45	39	34
13	14	56	50	43	38
13	15	61	54	47	42
13	16	65	59	51	45
13	17	70	63	55	49
13	18	75	67	59	53
13	19	80	72	63	57
13	20	84	76	67	60
13	21	89	80	71	64
13	22	94	85	75	68
13	23	98	89	79	72
13	24	103	94	83	75
13	25	108	98	87	79
14	14	61	55	47	42
14	15	66	59	51	46
14	16	71	64	56	50
14	17	77	69	60	54
14	18	82	74	65	58
14	19	87	78	69	63
14	20	92	83	73	67
14	21	97	88	78	71
14	22	102	93	82	75
14	23	107	98	87	79
14	24	113	102	91	83
14	25	118	107	95	87
15	15	72	64	56	51
15	16	77	70	61	55
15	17	83	75	66	60
15	18	88	80	70	64
15	19	94	85	75	69
15	20	100	90	80	73
15	21	105	96	85	78
15	22	111	101	90	82
15	23	116	106	94	87
15	24	122	111	99	91
15	25	128	117	104	96
16	16	83	75	66	60
16	17	89	81	71	65
16	18	95	86	76	70
16	19	101	92	82	74
16	20	107	98	87	79
16	21	113	103	92	84
16	22	119	109	97	89
16	23	125	115	102	94
16	24	131	120	108	99
16	25	137	126	113	104
17	17	96	87	77	70
17	18	102	93	82	75
17	19	109	99	88	81
17	20	115	105	93	86
17	21	121	111	99	91
17	22	128	117	105	96
17	23	134	123	110	102
17	24	141	129	116	107
17	25	147	135	122	112
18	18	109	99	88	81
18	19	116	106	94	87
18	20	123	112	100	92
18	21	130	119	106	98
18	22	136	125	112	104

n_1	n_2	5% / 10%	2½% / 5%	1% / 2%	½% / 1%
18	23	143	132	118	109
18	24	150	138	124	115
18	25	157	145	130	121
19	19	123	113	101	93
19	20	130	119	107	99
19	21	138	126	113	105
19	22	145	133	120	111
19	23	152	140	126	117
19	24	160	147	133	123
19	25	167	154	139	129
20	20	138	127	114	105
20	21	146	134	121	112
20	22	154	141	127	118
20	23	161	149	134	125
20	24	169	156	141	131
20	25	177	163	148	138
21	21	154	142	128	118
21	22	162	150	135	125
21	23	170	157	142	132
21	24	179	165	150	139
21	25	187	173	157	146
22	22	171	158	143	133
22	23	179	166	150	140
22	24	188	174	158	147
22	25	197	182	166	155
23	23	189	175	158	148
23	24	198	183	167	155
23	25	207	192	175	163
24	24	207	192	175	164
24	25	217	201	184	172
25	25	227	211	192	180
26	26	247	230	211	198
27	27	268	250	230	216
28	28	291	272	250	235
29	29	314	294	271	255
30	30	338	317	293	276
31	31	363	341	315	298
32	32	388	365	339	321
33	33	415	391	363	344
34	34	443	418	388	369
35	35	471	445	414	394
36	36	501	473	441	420
37	37	531	503	469	447
38	38	563	533	498	475
39	39	595	564	528	504
40	40	628	596	558	533
41	41	662	628	590	564
42	42	697	662	622	595
43	43	733	697	655	627
44	44	770	732	689	660
45	45	808	769	724	694
46	46	846	806	760	729
47	47	886	845	797	765
48	48	926	884	835	802
49	49	968	924	873	839
50	50	1010	965	913	877

For description, see page 28; for larger sample sizes, see page 35.

α_1	5%	2½%	1%	½%
α_2	10%	5%	2%	1%
$n_1\ n_2$				
2 2	—	—	—	—
2 3	—	—	—	—
2 4	—	—	—	—
2 5	10	—	—	—
2 6	12	—	—	—
2 7	14	—	—	—
2 8	16	16	—	—
2 9	18	18	—	—
2 10	18	20	—	—
2 11	20	22	—	—
2 12	22	24	—	—
2 13	24	26	26	—
2 14	24	26	28	—
2 15	26	28	30	—
2 16	28	30	32	—
2 17	30	32	34	—
2 18	32	34	36	—
2 19	32	36	38	38
2 20	34	38	40	40
2 21	36	38	42	42
2 22	38	40	44	44
2 23	38	42	44	46
2 24	40	44	46	48
2 25	42	46	48	50
3 3	9	—	—	—
3 4	12	—	—	—
3 5	15	15	—	—
3 6	15	18	—	—
3 7	18	21	21	—
3 8	21	21	24	—
3 9	21	24	27	27
3 10	24	27	30	30
3 11	27	30	33	33
3 12	27	30	33	36
3 13	30	33	36	39
3 14	33	36	39	42
3 15	33	36	42	42
3 16	36	39	45	45
3 17	36	42	45	48
3 18	39	45	48	51
3 19	42	45	51	54
3 20	42	48	54	57
3 21	45	51	54	57
3 22	48	51	57	60
3 23	48	54	60	63
3 24	51	57	63	66
3 25	54	60	66	69
4 4	16	16	—	—
4 5	16	20	20	—
4 6	18	20	24	24
4 7	21	24	28	28
4 8	24	28	32	32
4 9	27	28	32	36
4 10	28	30	36	36
4 11	29	33	40	40
4 12	36	36	40	44
4 13	35	39	44	48
4 14	38	42	48	48
4 15	40	44	48	52
4 16	44	48	52	56
4 17	44	48	56	60
4 18	46	50	56	60
4 19	49	53	57	64
4 20	52	60	64	68
4 21	52	59	64	72
4 22	56	62	66	72
4 23	57	64	69	76
4 24	60	68	76	80
4 25	63	68	75	84

α_1	5%	2½%	1%	½%
α_2	10%	5%	2%	1%
$n_1\ n_2$				
5 5	20	25	25	25
5 6	24	24	30	30
5 7	25	28	30	35
5 8	27	30	35	35
5 9	30	35	36	40
5 10	35	40	40	45
5 11	35	39	44	45
5 12	36	43	48	50
5 13	40	45	50	52
5 14	42	46	51	56
5 15	50	55	60	60
5 16	48	54	59	64
5 17	50	55	63	68
5 18	52	60	65	70
5 19	56	61	70	71
5 20	60	65	75	80
5 21	60	69	75	80
5 22	63	70	78	83
5 23	65	72	82	87
5 24	67	76	85	90
5 25	75	80	90	95
6 6	30	30	36	36
6 7	28	30	35	36
6 8	30	34	40	40
6 9	33	39	42	45
6 10	36	40	44	48
6 11	38	43	49	54
6 12	48	48	54	60
6 13	46	52	54	60
6 14	48	54	60	64
6 15	51	57	63	69
6 16	54	60	66	72
6 17	56	62	68	73
6 18	66	72	78	84
6 19	64	70	77	83
6 20	66	72	80	88
6 21	69	75	84	90
6 22	70	78	88	92
6 23	73	80	91	97
6 24	78	90	96	102
6 25	78	88	96	107
7 7	35	42	42	42
7 8	34	40	42	48
7 9	36	42	47	49
7 10	40	46	50	53
7 11	44	48	55	59
7 12	46	53	58	60
7 13	50	56	63	65
7 14	56	63	70	77
7 15	56	62	70	75
7 16	59	64	73	77
7 17	61	68	77	84
7 18	65	72	83	87
7 19	69	76	86	91
7 20	72	79	91	93
7 21	77	91	98	105
7 22	77	84	97	103
7 23	80	89	101	108
7 24	84	92	105	112
7 25	86	97	108	115
8 8	40	48	48	56
8 9	40	46	54	55
8 10	44	48	56	60
8 11	48	53	61	64
8 12	52	60	64	68
8 13	54	62	67	72
8 14	58	64	72	76
8 15	60	67	75	81

α_1	5%	2½%	1%	½%
α_2	10%	5%	2%	1%
$n_1\ n_2$				
8 16	72	80	88	88
8 17	68	77	85	88
8 18	72	80	88	94
8 19	74	82	93	98
8 20	80	88	100	104
8 21	81	89	102	107
8 22	84	94	106	112
8 23	89	98	107	115
8 24	96	104	120	128
8 25	95	104	118	125
9 9	54	54	63	63
9 10	50	53	61	63
9 11	52	59	63	70
9 12	57	63	69	75
9 13	59	65	73	78
9 14	63	70	80	84
9 15	69	75	84	90
9 16	69	78	87	94
9 17	74	82	92	99
9 18	81	90	99	108
9 19	80	90	99	107
9 20	84	93	104	111
9 21	90	99	111	117
9 22	91	101	113	122
9 23	94	106	117	126
9 24	99	111	123	132
9 25	101	114	125	135
10 10	60	70	70	80
10 11	57	60	69	77
10 12	60	66	74	80
10 13	64	70	78	84
10 14	68	74	84	90
10 15	75	80	90	100
10 16	76	84	94	100
10 17	79	89	99	106
10 18	82	92	104	108
10 19	85	94	104	113
10 20	100	110	120	130
10 21	95	105	118	126
10 22	98	108	120	130
10 23	101	114	127	137
10 24	106	118	130	140
10 25	110	125	140	150
11 11	66	77	88	88
11 12	64	72	77	86
11 13	67	75	86	91
11 14	73	82	90	96
11 15	76	84	95	102
11 16	80	89	100	106
11 17	85	93	104	110
11 18	88	97	108	118
11 19	92	102	114	122
11 20	96	107	118	127
11 21	101	112	124	134
11 22	110	121	143	143
11 23	108	119	132	142
11 24	111	124	139	150
11 25	117	129	143	154
12 12	72	84	96	96
12 13	71	81	92	95
12 14	78	86	94	104
12 15	84	93	102	108
12 16	88	96	108	116
12 17	90	100	112	119
12 18	96	108	120	126
12 19	99	108	121	130
12 20	104	116	128	140

α_1	5%	2½%	1%	½%
α_2	10%	5%	2%	1%
$n_1\ n_2$				
12 21	108	120	132	141
12 22	110	124	138	148
12 23	113	125	138	149
12 24	132	144	156	168
12 25	120	138	153	165
13 13	91	91	104	117
13 14	78	89	102	104
13 15	87	96	107	115
13 16	91	101	112	121
13 17	96	105	118	127
13 18	99	110	123	131
13 19	104	114	130	138
13 20	108	120	135	143
13 21	113	126	140	150
13 22	117	130	143	156
13 23	120	135	152	161
13 24	125	140	155	166
13 25	131	145	160	172
14 14	98	112	112	126
14 15	92	98	111	123
14 16	96	106	120	126
14 17	100	111	125	134
14 18	104	116	130	140
14 19	110	121	135	148
14 20	114	126	143	152
14 21	126	140	154	161
14 22	124	138	152	164
14 23	127	142	159	170
14 24	132	146	164	176
14 25	136	150	169	182
15 15	105	120	135	135
15 16	101	114	120	133
15 17	105	116	131	142
15 18	111	123	138	147
15 19	114	127	142	152
15 20	125	135	150	160
15 21	126	138	156	168
15 22	130	144	160	173
15 23	134	149	165	179
15 24	141	156	174	186
15 25	145	160	180	195
16 16	112	128	144	160
16 17	109	124	139	143
16 18	116	128	142	154
16 19	120	133	151	160
16 20	128	140	156	168
16 21	130	145	162	173
16 22	136	150	168	180
16 23	141	157	175	187
16 24	152	168	184	200
16 25	149	167	186	199
17 17	136	136	153	170
17 18	118	133	150	164
17 19	126	141	158	166
17 20	130	146	163	175
17 21	136	151	168	180
17 22	142	157	176	187
17 23	146	163	181	196
17 24	151	168	187	203
17 25	156	173	196	207
18 18	144	162	180	180
18 19	133	142	160	176
18 20	136	152	170	182
18 21	144	159	177	189
18 22	148	164	184	196

α_1	5%	2½%	1%	½%
α_2	10%	5%	2%	1%
$n_1\ n_2$				
18 23	152	170	189	204
18 24	162	180	198	216
18 25	162	180	202	216
19 19	152	171	190	190
19 20	144	160	171	187
19 21	147	163	184	199
19 22	152	169	190	204
19 23	159	177	197	209
19 24	164	183	204	218
19 25	168	187	211	224
20 20	160	180	200	220
20 21	154	173	193	199
20 22	160	176	196	212
20 23	164	184	205	219
20 24	172	192	212	228
20 25	180	200	220	235
21 21	168	189	210	231
21 22	163	183	205	223
21 23	171	189	213	227
21 24	177	198	222	237
21 25	182	202	225	244
22 22	190	198	242	242
22 23	173	194	217	237
22 24	182	204	228	242
22 25	189	209	234	250
23 23	207	230	253	253
23 24	183	205	228	249
23 25	195	216	243	262
24 24	216	240	264	288
24 25	204	225	254	262
25 25	225	250	275	300
26 26	234	260	286	312
27 27	243	270	324	324
28 28	280	308	336	364
29 29	290	319	348	377
30 30	300	330	360	390
31 31	310	341	372	403
32 32	320	352	416	416
33 33	330	396	429	462
34 34	374	408	442	476
35 35	385	420	455	490
36 36	396	432	468	504
37 37	407	444	518	518
38 38	418	456	532	570
39 39	429	468	546	585
40 40	440	520	560	600
41 41	492	533	574	615
42 42	504	546	588	630
43 43	516	559	645	688
44 44	528	572	660	704
45 45	540	585	675	720
46 46	552	644	690	736
47 47	564	658	705	752
48 48	576	672	720	768
49 49	637	686	735	833
50 50	650	700	800	850

31

For description, see page 28; for larger sample sizes, see page 35.

Critical values for the Kruskal-Wallis test (small sample sizes)

$$H = \frac{12}{N(N+1)} \sum_{i=1}^{k} \frac{R_i^2}{n_i} - 3(N+1)$$

k = 3 samples (N ≤ 19)

sample sizes			α 10%	5%	2%	1%
1	1	1	–	–	–	–
2	1	1	–	–	–	–
2	2	1	–	–	–	–
2	2	2	4.571	–	–	–
3	1	1	–	–	–	–
3	2	1	4.286	–	–	–
3	2	2	4.500	4.714	–	–
3	3	1	4.571	5.143	–	–
3	3	2	4.556	5.361	6.250	–
3	3	3	4.622	5.600	6.489	7.200
4	1	1	–	–	–	–
4	2	1	4.500	–	–	–
4	2	2	4.458	5.333	6.000	–
4	3	1	4.056	5.208	–	–
4	3	2	4.511	5.444	6.144	6.444
4	3	3	4.709	5.791	6.564	6.745
4	4	1	4.167	4.967	6.667	6.667
4	4	2	4.555	5.455	6.600	7.036
4	4	3	4.545	5.598	6.712	7.144
4	4	4	4.654	5.692	6.962	7.654
5	1	1	–	–	–	–
5	2	1	4.200	5.000	–	–
5	2	2	4.373	5.160	6.000	6.533
5	3	1	4.018	4.960	6.044	–
5	3	2	4.651	5.251	6.124	6.909
5	3	3	4.533	5.648	6.533	7.079
5	4	1	3.987	4.985	6.431	6.955
5	4	2	4.541	5.273	6.505	7.205
5	4	3	4.549	5.656	6.676	7.445
5	4	4	4.668	5.657	6.953	7.760
5	5	1	4.109	5.127	6.145	7.309
5	5	2	4.623	5.338	6.446	7.338
5	5	3	4.545	5.705	6.866	7.578
5	5	4	4.523	5.666	7.000	7.823
5	5	5	4.560	5.780	7.220	8.000
6	1	1	–	–	–	–
6	2	1	4.200	4.822	–	–
6	2	2	4.545	5.345	6.182	6.655
6	3	1	3.909	4.855	6.236	6.873
6	3	2	4.682	5.348	6.227	6.970
6	3	3	4.590	5.615	6.590	7.410
6	4	1	4.038	4.947	6.174	7.106
6	4	2	4.494	5.340	6.571	7.340
6	4	3	4.604	5.610	6.725	7.500
6	4	4	4.595	5.681	6.900	7.795
6	5	1	4.128	4.990	6.138	7.182
6	5	2	4.596	5.338	6.585	7.376
6	5	3	4.535	5.602	6.829	7.590
6	5	4	4.522	5.661	7.018	7.936
6	5	5	4.547	5.729	7.110	8.028
6	6	1	4.000	4.945	6.286	7.121
6	6	2	4.438	5.410	6.667	7.467
6	6	3	4.558	5.625	6.900	7.725
6	6	4	4.548	5.724	7.107	8.000
6	6	5	4.542	5.765	7.152	8.124
6	6	6	4.643	5.801	7.240	8.222
7	1	1	4.267	–	–	–
7	2	1	4.200	4.706	5.891	–
7	2	2	4.526	5.143	6.058	7.000
7	3	1	4.173	4.952	6.043	7.030
7	3	2	4.582	5.357	6.339	6.839
7	3	3	4.603	5.620	6.656	7.228
7	4	1	4.121	4.986	6.319	6.986
7	4	2	4.549	5.376	6.447	7.321
7	4	3	4.527	5.623	6.780	7.550
7	4	4	4.562	5.650	6.962	7.814
7	5	1	4.035	5.064	6.194	7.061
7	5	2	4.485	5.393	6.477	7.450
7	5	3	4.535	5.607	6.874	7.697
7	5	4	4.542	5.733	7.084	7.931
7	5	5	4.571	5.708	7.101	8.108
7	6	1	4.033	5.067	6.214	7.254
7	6	2	4.500	5.357	6.587	7.490
7	6	3	4.550	5.689	6.930	7.756
7	6	4	4.562	5.706	7.086	8.039
7	6	5	4.560	5.770	7.191	8.157
7	6	6	4.530	5.730	7.197	8.257
7	7	1	3.986	4.986	6.300	7.157
7	7	2	4.491	5.398	6.693	7.491
7	7	3	4.613	5.688	7.003	7.810
7	7	4	4.563	5.766	7.145	8.142
7	7	5	4.546	5.746	7.247	8.257
8	1	1	4.418	–	–	–
8	2	1	4.011	4.909	6.000	–
8	2	2	4.587	5.356	5.962	6.663
8	3	1	4.010	4.881	6.179	6.804
8	3	2	4.451	5.316	6.371	7.022
8	3	3	4.543	5.617	6.683	7.350
8	4	1	4.038	5.044	6.140	6.973
8	4	2	4.500	5.393	6.536	7.350
8	4	3	4.529	5.623	6.854	7.585
8	4	4	4.561	5.779	7.075	7.853
8	5	1	3.967	4.869	6.257	7.110
8	5	2	4.466	5.415	6.571	7.440
8	5	3	4.514	5.614	6.932	7.706
8	5	4	4.549	5.718	7.051	7.992
8	5	5	4.555	5.769	7.159	8.116
8	6	1	4.015	5.015	6.358	7.256
8	6	2	4.463	5.404	6.618	7.522
8	6	3	4.575	5.678	6.980	7.796
8	6	4	4.563	5.743	7.120	8.045
8	6	5	4.550	5.750	7.221	8.226
8	7	1	4.045	5.041	6.366	7.308
8	7	2	4.451	5.403	6.619	7.571
8	7	3	4.556	5.698	7.021	7.827
8	7	4	4.548	5.759	7.153	8.118
8	8	1	4.044	5.039	6.294	7.314
8	8	2	4.509	5.408	6.711	7.654
8	8	3	4.555	5.734	7.021	7.889
9	1	1	4.545	–	–	–
9	2	1	3.906	4.842	5.662	6.346
9	2	2	4.484	5.260	6.095	6.897
9	3	1	4.073	4.952	6.095	6.886
9	3	2	4.492	5.340	6.359	7.006
9	3	3	4.633	5.589	6.800	7.422
9	4	1	3.971	5.071	6.130	7.171
9	4	2	4.489	5.400	6.518	7.364
9	4	3	4.526	5.652	6.882	7.614
9	4	4	4.576	5.704	6.990	7.910
9	5	1	4.056	5.040	6.349	7.149
9	5	2	4.465	5.396	6.596	7.447
9	5	3	4.587	5.670	6.972	7.733
9	5	4	4.531	5.713	7.121	8.025
9	5	5	4.557	5.770	7.213	8.173
9	6	1	3.953	5.049	6.255	7.248
9	6	2	4.481	5.392	6.614	7.566
9	6	3	4.548	5.671	6.975	7.823
9	6	4	4.546	5.745	7.130	8.109
9	7	1	4.011	5.042	6.397	7.282
9	7	2	4.480	5.429	6.679	7.637
9	7	3	4.547	5.656	6.998	7.861
9	8	1	3.986	4.985	6.351	7.394
9	8	2	4.492	5.420	6.679	7.642
9	9	1	4.007	4.961	6.407	7.333
10	1	1	4.654	4.654	–	–
10	2	1	4.114	4.840	5.776	6.429
10	2	2	4.434	5.120	6.034	6.537
10	3	1	3.996	5.076	6.053	6.851
10	3	2	4.470	5.362	6.375	7.042
10	3	3	4.529	5.588	6.784	7.372
10	4	1	4.042	5.018	6.158	7.105
10	4	2	4.462	5.345	6.492	7.357
10	4	3	4.588	5.661	6.905	7.617
10	4	4	4.565	5.716	7.065	7.907
10	5	1	3.988	4.954	6.318	7.178
10	5	2	4.455	5.420	6.612	7.514
10	5	3	4.552	5.636	6.938	7.752
10	5	4	4.557	5.744	7.135	8.048
10	6	1	3.967	5.042	6.383	7.316
10	6	2	4.480	5.406	6.669	7.588
10	6	3	4.551	5.656	7.002	7.882
10	7	1	3.981	4.986	6.370	7.252
10	7	2	4.492	5.377	6.652	7.641
10	8	1	3.964	5.038	6.414	7.359
11	1	1	4.028	4.747	–	–
11	2	1	4.044	4.816	5.834	6.600
11	2	2	4.414	5.164	6.050	6.766
11	3	1	3.985	5.030	6.030	6.818
11	3	2	4.487	5.374	6.379	7.094
11	3	3	4.589	5.583	6.776	7.418
11	4	1	3.991	4.988	6.111	7.090
11	4	2	4.484	5.365	6.553	7.396
11	4	3	4.536	5.660	6.881	7.679
11	4	4	4.550	5.740	7.036	7.945
11	5	1	4.026	5.020	6.284	7.130
11	5	2	4.490	5.374	6.648	7.507
11	5	3	4.550	5.646	6.962	7.807
11	6	1	4.029	5.062	6.304	7.261
11	6	2	4.463	5.408	6.693	7.564
11	7	1	4.045	4.985	6.409	7.330
12	1	1	4.148	4.829	–	–
12	2	1	4.092	4.875	5.550	6.229
12	2	2	4.379	5.173	5.967	6.761
12	3	1	3.930	4.930	6.018	6.812
12	3	2	4.477	5.350	6.412	7.134
12	3	3	4.579	5.576	6.746	7.471
12	4	1	4.003	4.931	6.225	7.108
12	4	2	4.500	5.442	6.547	7.389
12	4	3	4.524	5.661	6.903	7.703
12	5	1	3.985	4.977	6.326	7.215
12	5	2	4.486	5.395	6.649	7.512
12	6	1	4.050	5.005	6.371	7.297
13	1	1	4.254	4.900	–	–
13	2	1	3.989	4.819	5.727	6.312
13	2	2	4.385	5.199	6.134	6.792
13	3	1	4.095	5.024	6.081	6.846
13	3	2	4.485	5.371	6.407	7.138
13	3	3	4.539	5.613	6.755	7.449
13	4	1	4.045	4.963	6.325	7.052
13	4	2	4.484	5.368	6.587	7.434
13	5	1	4.043	4.993	6.288	7.238
14	1	1	3.728	4.963	–	–
14	2	1	4.070	4.863	5.737	6.356
14	2	2	4.441	5.193	6.045	6.812
14	3	1	4.075	4.977	6.029	6.811
14	3	2	4.515	5.383	6.413	7.218
14	4	1	4.020	4.991	6.265	7.176
15	1	1	3.843	5.020	–	–
15	2	1	4.032	4.827	5.599	6.053
15	2	2	4.461	5.184	6.044	6.760
15	3	1	4.055	5.019	6.139	6.813
16	1	1	3.886	4.511	5.070	–
16	2	1	4.044	4.849	5.670	6.189
17	1	1	3.986	4.581	5.116	–
∞	∞	∞	4.605	5.991	7.824	9.210

32

For description, see page 28; for larger equal sample sizes, see page 34.

k = 4 samples (N ≤ 14)

sample sizes				α 10%	5%	2%	1%
1	1	1	1	–	–	–	–
2	1	1	1	–	–	–	–
2	2	1	1	–	–	–	–
2	2	2	1	5.357	5.679	–	–
2	2	2	2	5.667	6.167	6.667	6.667
3	1	1	1	–	–	–	–
3	2	1	1	5.143	–	–	–
3	2	2	1	5.556	5.833	6.500	–
3	2	2	2	5.644	6.333	6.978	7.133
3	3	1	1	5.333	6.333	–	–
3	3	2	1	5.689	6.244	6.689	7.200
3	3	2	2	5.745	6.527	7.182	7.636
3	3	3	1	5.655	6.600	7.109	7.400
3	3	3	2	5.879	6.727	7.636	8.015
3	3	3	3	6.026	7.000	7.872	8.538
4	1	1	1	–	–	–	–
4	2	1	1	5.250	5.833	–	–
4	2	2	1	5.533	6.133	6.667	7.000
4	2	2	2	5.755	6.545	7.091	7.391
4	3	1	1	5.067	6.178	6.711	7.067
4	3	2	1	5.591	6.309	7.018	7.455
4	3	2	2	5.750	6.621	7.530	7.871
4	3	3	1	5.689	6.545	7.485	7.758
4	3	3	2	5.872	6.795	7.763	8.333
4	3	3	3	6.016	6.984	7.995	8.659
4	4	1	1	5.182	5.945	7.091	7.909
4	4	2	1	5.568	6.386	7.364	7.909
4	4	2	2	5.808	6.731	7.750	8.346
4	4	3	1	5.692	6.635	7.660	8.231
4	4	3	2	5.901	6.874	7.951	8.621
4	4	3	3	6.019	7.038	8.181	8.876
4	4	4	1	5.654	6.725	7.879	8.588
4	4	4	2	5.914	6.957	8.157	8.871
5	1	1	1	5.333	–	–	–
5	2	1	1	5.267	5.960	6.600	–
5	2	2	1	5.542	6.109	6.927	7.276
5	2	2	2	5.636	6.564	7.364	7.773
5	3	1	1	5.160	6.004	6.964	7.400
5	3	2	1	5.518	6.364	7.285	7.758
5	3	2	2	5.772	6.664	7.626	8.203
5	3	3	1	5.667	6.641	7.656	8.128
5	3	3	2	5.866	6.822	7.912	8.607
5	3	3	3	6.021	7.019	8.124	8.848
5	4	1	1	5.255	6.041	7.182	7.909
5	4	2	1	5.581	6.419	7.477	8.173
5	4	2	2	5.782	6.725	7.849	8.473
5	4	3	1	5.656	6.685	7.793	8.409
5	4	3	2	5.902	6.926	8.069	8.802
5	4	4	1	5.674	6.760	7.986	8.726
5	5	1	1	5.154	6.077	7.308	8.108
5	5	2	1	5.585	6.541	7.536	8.327
5	5	2	2	5.800	6.777	7.943	8.634
5	5	3	1	5.663	6.745	7.857	8.611
6	1	1	1	5.667	5.667	–	–
6	2	1	1	5.145	5.964	6.600	7.036
6	2	2	1	5.470	6.242	7.000	7.500
6	2	2	2	5.744	6.538	7.513	7.923
6	3	1	1	5.197	6.045	7.091	7.621
6	3	2	1	5.577	6.397	7.321	7.885
6	3	2	2	5.780	6.703	7.758	8.262
6	3	3	1	5.659	6.637	7.725	8.220
6	3	3	2	5.886	6.876	7.962	8.695
6	4	1	1	5.186	6.071	7.250	8.000
6	4	2	1	5.571	6.489	7.516	8.302
6	4	2	2	5.810	6.743	7.929	8.610
6	4	3	1	5.637	6.710	7.915	8.530
6	5	1	1	5.176	6.110	7.218	8.141
6	5	2	1	5.589	6.541	7.598	8.389
6	6	1	1	5.219	6.133	7.276	8.181
7	1	1	1	5.197	5.945	–	–
7	2	1	1	5.097	6.006	6.786	7.273
7	2	2	1	5.484	6.319	7.011	7.626
7	2	2	2	5.689	6.565	7.568	8.053
7	3	1	1	5.147	6.070	7.037	7.652
7	3	2	1	5.576	6.466	7.383	8.005
7	3	2	2	5.795	6.718	7.759	8.407
7	3	3	1	5.664	6.671	7.721	8.352
7	4	1	1	5.169	6.104	7.222	8.032
7	4	2	1	5.580	6.543	7.531	8.337
7	5	1	1	5.225	6.113	7.318	8.148
8	1	1	1	4.955	6.182	–	–
8	2	1	1	5.154	5.933	6.692	7.423
8	2	2	1	5.481	6.305	7.096	7.648
8	2	2	2	5.714	6.571	7.600	8.207
8	3	1	1	5.231	6.099	7.154	7.788
8	3	2	1	5.579	6.464	7.455	8.114
8	4	1	1	5.200	6.143	7.271	8.029
9	1	1	1	4.880	5.701	6.385	–
9	2	1	1	5.128	5.919	6.725	7.326
9	2	2	1	5.492	6.292	7.130	7.692
9	3	1	1	5.216	6.105	7.171	7.768
10	1	1	1	5.037	5.908	6.560	–
10	2	1	1	5.194	5.937	6.794	7.251
11	1	1	1	4.969	5.457	6.714	–
∞	∞	∞	∞	6.251	7.815	9.837	11.34

k = 5 samples (N ≤ 13)

sample sizes					α 10%	5%	2%	1%
1	1	1	1	1		–	–	
2	1	1	1	1	–	–	–	–
2	2	1	1	1	5.786	–	–	–
2	2	2	1	1	6.250	6.750	–	–
2	2	2	2	1	6.600	7.133	7.533	7.533
2	2	2	2	2	6.982	7.418	8.073	8.291
3	1	1	1	1	–	–	–	–
3	2	1	1	1	6.139	6.583	–	–
3	2	2	1	1	6.511	6.800	7.400	7.600
3	2	2	2	1	6.709	7.309	7.836	8.127
3	2	2	2	2	6.955	7.682	8.303	8.682
3	3	1	1	1	6.311	7.111	7.467	–
3	3	2	1	1	6.600	7.200	7.782	8.073
3	3	2	2	1	6.788	7.591	8.258	8.576
3	3	2	2	2	7.026	7.910	8.667	9.115
3	3	3	1	1	6.788	7.576	8.242	8.424
3	3	3	2	1	6.910	7.769	8.590	9.051
3	3	3	2	2	7.121	8.044	9.011	9.505
3	3	3	3	1	7.077	8.000	8.879	9.451
4	1	1	1	1	6.167	–	–	–
4	2	1	1	1	6.200	6.733	7.267	–
4	2	2	1	1	6.491	7.145	7.636	8.036
4	2	2	2	1	6.773	7.500	8.205	8.545
4	2	2	2	2	7.000	7.846	8.673	9.077
4	3	1	1	1	6.227	7.073	7.691	8.236
4	3	2	1	1	6.614	7.439	8.091	8.394
4	3	2	2	1	6.833	7.679	8.545	8.962
4	3	2	2	2	7.055	7.984	8.918	9.429
4	3	3	1	1	6.737	7.660	8.513	8.891
4	3	3	2	1	6.956	7.874	8.830	9.374
4	4	1	1	1	6.364	7.114	8.182	8.636
4	4	2	1	1	6.654	7.500	8.385	8.885
4	4	2	2	1	6.890	7.797	8.802	9.330
4	4	3	1	1	6.758	7.714	8.742	9.247
5	1	1	1	1	6.667	6.667	–	–
5	2	1	1	1	6.295	6.905	7.418	7.855
5	2	2	1	1	6.527	7.273	7.909	8.318
5	2	2	2	1	6.754	7.600	8.408	8.831
5	2	2	2	2	6.989	7.925	8.782	9.316
5	3	1	1	1	6.364	7.164	7.939	8.303
5	3	2	1	1	6.641	7.462	8.272	8.756
5	3	2	2	1	6.866	7.756	8.705	9.251
5	3	3	1	1	6.725	7.684	8.651	9.187
5	4	1	1	1	6.388	7.154	8.236	8.831
5	4	2	1	1	6.679	7.520	8.492	9.152
5	5	1	1	1	6.356	7.226	8.334	9.152
6	1	1	1	1	6.073	7.091	–	–
6	2	1	1	1	6.288	6.909	7.682	8.000
6	2	2	1	1	6.577	7.308	8.051	8.628
6	2	2	2	1	6.802	7.593	8.549	9.077
6	3	1	1	1	6.423	7.051	8.064	8.590
6	3	2	1	1	6.648	7.505	8.407	9.000
6	4	1	1	1	6.396	7.187	8.286	9.033
7	1	1	1	1	6.182	6.831	7.455	–
7	2	1	1	1	6.368	6.984	7.753	8.231
7	2	2	1	1	6.593	7.356	8.143	8.689
7	3	1	1	1	6.367	7.152	8.119	8.779
8	1	1	1	1	6.087	6.538	7.769	–
8	2	1	1	1	6.338	6.997	7.821	8.308
9	1	1	1	1	6.095	6.755	7.458	8.044
∞	∞	∞	∞	∞	7.779	9.488	11.67	13.28

k = 6 samples (N ≤ 13)

sample sizes						α 10%	5%	2%	1%
1	1	1	1	1	1	–	–	–	–
2	1	1	1	1	1	–	–	–	–
2	2	1	1	1	1	6.833	–	–	–
2	2	2	1	1	1	7.267	7.600	7.800	–
2	2	2	2	1	1	7.527	8.018	8.455	8.618
2	2	2	2	2	1	7.909	8.455	9.000	9.227
2	2	2	2	2	2	8.154	8.846	9.538	9.846
3	1	1	1	1	1	–	–	–	–
3	2	1	1	1	1	7.133	7.467	–	–
3	2	2	1	1	1	7.418	7.945	8.345	8.509
3	2	2	2	1	1	7.727	8.348	8.939	9.136
3	2	2	2	2	1	7.987	8.731	9.346	9.692
3	2	2	2	2	2	8.198	9.033	9.813	10.22
3	3	1	1	1	1	7.400	7.909	8.564	8.564
3	3	2	1	1	1	7.697	8.303	8.803	9.045
3	3	2	2	1	1	7.872	8.615	9.269	9.628
3	3	2	2	2	1	8.077	8.923	9.714	10.15
3	3	3	1	1	1	7.821	8.641	9.205	9.564
3	3	3	2	1	1	8.000	8.835	9.670	10.08
4	1	1	1	1	1	7.333	7.333	–	–
4	2	1	1	1	1	7.255	7.827	8.236	8.400
4	2	2	1	1	1	7.545	8.205	8.727	9.000
4	2	2	2	1	1	7.808	8.558	9.192	9.538
4	2	2	2	2	1	8.044	8.868	9.643	10.07
4	3	1	1	1	1	7.394	8.053	8.758	9.023
4	3	2	1	1	1	7.679	8.429	9.115	9.506
4	3	2	2	1	1	7.929	8.742	9.577	10.01
4	3	3	1	1	1	7.780	8.654	9.495	9.934
4	4	1	1	1	1	7.404	8.231	9.096	9.538
4	4	2	1	1	1	7.714	8.571	9.445	9.940
5	1	1	1	1	1	7.385	7.909	–	–
5	2	1	1	1	1	7.345	7.891	8.473	8.682
5	2	2	1	1	1	7.638	8.308	8.938	9.362
5	2	2	1	1	1	7.833	8.624	9.442	9.890
5	3	1	1	1	1	7.369	8.169	9.062	9.503
5	3	2	1	1	1	7.701	8.495	9.371	9.837
5	4	1	1	1	1	7.503	8.242	9.234	9.841
6	1	1	1	1	1	7.197	7.879	8.409	–
6	2	1	1	1	1	7.397	8.013	8.692	9.051
6	2	2	1	1	1	7.626	8.374	9.165	9.604
6	3	1	1	1	1	7.473	8.209	9.176	9.659
7	1	1	1	1	1	7.198	7.791	8.846	8.846
7	2	1	1	1	1	7.389	8.119	8.821	9.268
8	1	1	1	1	1	7.154	7.788	8.712	9.231
∞	∞	∞	∞	∞	∞	9.236	11.07	13.39	15.09

33

For description, see page 28; for larger equal sample sizes, see page 34.

Critical values for the Kruskal-Wallis test (equal sample sizes)

$$H = \frac{12}{n^2 k(nk + 1)} \sum_{i=1}^{k} R_i^2 - 3(nk + 1)$$

n	k=3 10%	k=3 5%	k=3 2%	k=3 1%	k=4 10%	k=4 5%	k=4 2%	k=4 1%	k=5 10%	k=5 5%	k=5 2%	k=5 1%	k=6 10%	k=6 5%	k=6 2%	k=6 1%
2	4.571	–	–	–	5.667	6.167	6.667	6.667	6.982	7.418	8.073	8.291	8.154	8.846	9.538	9.846
3	4.622	5.600	6.489	7.200	6.026	7.000	7.872	8.538	7.333	8.333	9.467	10.20	8.620	9.789	11.03	11.82
4	4.654	5.692	6.962	7.654	6.088	7.235	8.515	9.287	7.457	8.685	10.13	11.07	8.800	10.14	11.71	12.72
5	4.560	5.780	7.220	8.000	6.120	7.377	8.863	9.789	7.532	8.876	10.47	11.57	8.902	10.36	12.07	13.26
6	4.643	5.801	7.240	8.222	6.127	7.453	9.027	10.09	7.557	9.002	10.72	11.91	8.958	10.50	12.33	13.60
7	4.594	5.819	7.332	8.378	6.141	7.501	9.152	10.25	7.600	9.080	10.87	12.14	8.992	10.59	12.50	13.84
8	4.595	5.805	7.355	8.465	6.148	7.534	9.250	10.42	7.624	9.126	10.99	12.29	9.037	10.66	12.62	13.99
9	4.586	5.831	7.418	8.529	6.161	7.557	9.316	10.53	7.637	9.166	11.06	12.41	9.057	10.71	12.71	14.13
10	4.581	5.853	7.453	8.607	6.167	7.586	9.376	10.62	7.650	9.200	11.13	12.50	9.078	10.75	12.78	14.24
11	4.587	5.885	7.489	8.648	6.163	7.623	9.422	10.69	7.660	9.242	11.19	12.58	9.093	10.76	12.84	14.32
12	4.578	5.872	7.523	8.712	6.185	7.629	9.458	10.75	7.675	9.274	11.22	12.63	9.105	10.79	12.90	14.38
13	4.601	5.901	7.551	8.735	6.191	7.645	9.481	10.80	7.685	9.303	11.27	12.69	9.115	10.83	12.93	14.44
14	4.592	5.896	7.566	8.754	6.198	7.658	9.508	10.84	7.695	9.307	11.29	12.74	9.125	10.84	12.98	14.49
15	4.591	5.902	7.582	8.821	6.201	7.676	9.531	10.87	7.701	9.302	11.32	12.77	9.133	10.86	13.01	14.53
16	4.595	5.909	7.596	8.822	6.205	7.678	9.550	10.90	7.705	9.313	11.34	12.79	9.140	10.88	13.03	14.56
17	4.593	5.915	7.609	8.856	6.206	7.682	9.568	10.92	7.709	9.325	11.36	12.83	9.144	10.88	13.04	14.60
18	4.596	5.932	7.622	8.865	6.212	7.698	9.583	10.95	7.714	9.334	11.38	12.85	9.149	10.89	13.06	14.63
19	4.598	5.923	7.634	8.887	6.212	7.701	9.595	10.98	7.717	9.342	11.40	12.87	9.156	10.90	13.07	14.64
20	4.594	5.926	7.641	8.905	6.216	7.703	9.606	10.98	7.719	9.353	11.41	12.91	9.159	10.92	13.09	14.67
21	4.597	5.930	7.652	8.918	6.218	7.709	9.623	11.01	7.723	9.356	11.43	12.92	9.164	10.93	13.11	14.70
22	4.597	5.932	7.657	8.928	6.215	7.714	9.629	11.03	7.724	9.362	11.43	12.92	9.168	10.94	13.12	14.72
23	4.598	5.937	7.664	8.947	6.220	7.719	9.640	11.03	7.727	9.368	11.44	12.94	9.171	10.93	13.13	14.74
24	4.598	5.936	7.670	8.964	6.221	7.724	9.652	11.06	7.729	9.375	11.45	12.96	9.170	10.93	13.14	14.74
25	4.599	5.942	7.682	8.975	6.222	7.727	9.659	11.07	7.730	9.377	11.46	12.96	9.177	10.94	13.15	14.77
∞	4.605	5.991	7.824	9.210	6.251	7.815	9.837	11.34	7.779	9.488	11.67	13.28	9.236	11.07	13.39	15.09

For description, see page 28.

Critical values for Friedman's test

$$M = \frac{12}{nk(k + 1)} \sum_{i=1}^{k} R_i^2 - 3n(k + 1)$$

n	k=3 10%	k=3 5%	k=3 2%	k=3 1%	k=4 10%	k=4 5%	k=4 2%	k=4 1%	k=5 10%	k=5 5%	k=5 2%	k=5 1%	k=6 10%	k=6 5%	k=6 2%	k=6 1%
2	–	–	–	–	6.000	6.000	–	–	7.200	7.600	8.000	8.000	8.286	9.143	9.429	9.714
3	6.000	6.000	–	–	6.600	7.400	8.200	9.000	7.467	8.533	9.600	10.13	8.714	9.857	11.00	11.76
4	6.000	6.500	8.000	8.000	6.300	7.800	8.400	9.600	7.600	8.800	10.20	11.20	9.000	10.29	11.71	12.71
5	5.200	6.400	8.400	8.400	6.360	7.800	9.000	9.960	7.680	8.960	10.56	11.68	9.000	10.49	12.09	13.23
6	5.333	7.000	8.333	9.000	6.400	7.600	9.400	10.20	7.733	9.067	10.80	11.87	9.048	10.57	12.38	13.62
7	5.429	7.143	8.000	8.857	6.429	7.800	9.171	10.54	7.771	9.143	10.97	12.11	9.122	10.67	12.55	13.86
8	5.250	6.250	7.750	9.000	6.300	7.650	9.450	10.50	7.700	9.200	11.00	12.30	9.071	10.71	12.64	14.00
9	5.556	6.222	8.000	9.556	6.200	7.667	9.400	10.73	7.733	9.244	11.11	12.44	9.127	10.78	12.75	14.14
10	5.000	6.200	7.800	9.600	6.360	7.680	9.480	10.68	7.760	9.280	11.20	12.48	9.143	10.80	12.80	14.23
11	5.091	6.545	7.818	9.455	6.273	7.691	9.655	10.75	7.782	9.309	11.20	12.58	9.130	10.84	12.92	14.32
12	5.167	6.500	8.000	9.500	6.300	7.700	9.500	10.80	7.733	9.333	11.27	12.60	9.143	10.86	12.95	14.38
13	4.769	6.615	8.000	9.385	6.138	7.800	9.646	10.85	7.754	9.354	11.32	12.68	9.176	10.89	13.00	14.45
14	5.143	6.143	8.143	9.143	6.343	7.714	9.600	10.89	7.771	9.371	11.37	12.74	9.184	10.90	13.02	14.49
15	4.933	6.400	8.133	8.933	6.280	7.720	9.640	10.92	7.787	9.387	11.36	12.80	9.210	10.92	13.06	14.54
16	4.875	6.500	7.875	9.375	6.300	7.800	9.600	10.95	7.750	9.400	11.40	12.80	9.214	10.96	13.07	14.57
17	5.059	6.118	7.529	9.294	6.318	7.800	9.635	11.05	7.765	9.412	11.44	12.85	9.202	10.95	13.10	14.61
18	4.778	6.333	8.111	9.000	6.333	7.733	9.667	10.93	7.778	9.422	11.47	12.89	9.206	10.95	13.11	14.63
19	5.053	6.421	7.895	9.579	6.347	7.863	9.632	11.02	7.789	9.432	11.45	12.88	9.196	11.00	13.14	14.67
20	4.900	6.300	7.900	9.300	6.240	7.800	9.600	11.10	7.760	9.400	11.48	12.92	9.200	11.00	13.11	14.66
21	4.952	6.095	7.714	9.238	6.314	7.800	9.686	11.06	7.771	9.448	11.50	12.91	9.218	10.99	13.14	14.69
22	4.727	6.091	8.273	9.091	6.327	7.800	9.709	11.07	7.782	9.418	11.49	12.95	9.221	10.96	13.14	14.73
23	4.957	6.348	8.087	9.391	6.287	7.800	9.678	11.09	7.791	9.426	11.51	12.97	9.236	11.00	13.19	14.73
24	5.083	6.250	7.750	9.250	6.250	7.750	9.700	11.15	7.767	9.433	11.50	13.00	9.238	10.95	13.19	14.74
25	4.880	6.080	7.760	8.960	6.264	7.800	9.672	11.16	7.776	9.440	11.52	12.99	9.229	10.99	13.21	14.74
∞	4.605	5.991	7.824	9.210	6.251	7.815	9.837	11.34	7.779	9.488	11.67	13.28	9.236	11.07	13.39	15.09

For description, see pages 28–9.

Critical values for nonparametric tests with large samples

For all the eight tests dealt with on pages 26–34 there are approximate methods for finding critical values when sample sizes exceed those covered in the tables.

Approximate critical values for the **sign test, Wilcoxon signed-rank test** and **Mann–Whitney U test** may be found from the table of percentage points of the standard normal distribution on page 20. Denote by z the appropriate percentage point of the standard normal distribution, e.g. 1.9600 for an $\alpha_2 = 5\%$ two-sided test or 1.6449 for an $\alpha_1 = 5\%$ one-sided test. Then calculate μ and σ from the table below. The required critical value is $[\mu - z\sigma - \frac{1}{2}]$, the square brackets denoting the integer part.

	μ	σ
sign test	$\frac{1}{2}n$	$\frac{1}{2}\sqrt{n}$
Wilcoxon signed-rank test	$\frac{1}{4}n(n+1)$	$\{\frac{1}{24}n(n+1)(2n+1)\}^{1/2}$
Mann–Whitney U test	$\frac{1}{2}n_1 n_2$	$\{\frac{1}{12}n_1 n_2(n_1 + n_2 + 1)\}^{1/2}$

For example in the sign test with sample size $n = 144$, $\mu = \frac{1}{2}(144) = 72$ and $\sigma = \frac{1}{2}\sqrt{144} = 6$, so that the $\alpha_2 = 5\%$ critical value is $[72 - 1.96 \times 6 - \frac{1}{2}] = [59.74] = 59$, i.e. the $\alpha_2 = 5\%$ critical region is $S \leqslant 59$. The reader may verify similarly that (i) for the signed-rank test with $n = 144$: $\mu = 5220$, $\sigma = 501.428$, and the $\alpha_2 = 5\%$ critical region is $T \leqslant 4236$; and (ii) in the Mann–Whitney test with sample sizes 25 and 30: $\mu = 375$, $\sigma = 59.161$, and the $\alpha_2 = 5\%$ critical region is $U \leqslant 258$.

For the **Kolmogorov–Smirnov goodness-of-fit test**, approximate critical values are simply found by dividing the constants b in the following table by $\sqrt{n}$:

α_1	5%	$2\frac{1}{2}$%	1%	$\frac{1}{2}$%
α_2	10%	5%	2%	1%
b	1.2238	1.3581	1.5174	1.6276
c	0.8255	0.8993	0.9885	1.0600

So with a sample of size $n = 144$, the $\alpha_2 = 5\%$ critical value is $1.3581/\sqrt{144} = 0.1132$, i.e. the critical region is $D_{144} \geqslant 0.1132$. The same constants b are used to obtain approximate critical regions for the **Kolmogorov–Smirnov two-sample test**. In this case b is multiplied by $\{1/n_1 + 1/n_2\}^{1/2}$ to give critical values for D (not D^*). So with sample sizes 25 and 30, $\{1/n_1 + 1/n_2\}^{1/2} = \{1/25 + 1/30\}^{1/2} = 0.2708$ and the $\alpha_2 = 5\%$ critical region is $D \geqslant 1.3581 \times 0.2708 = 0.3678$. For the **Kolmogorov–Smirnov test for normality** (with unspecified mean and standard deviation), the critical values are found as in the goodness-of-fit test except that the second row of constants c is used instead of b. In this case the $\alpha_2 = 5\%$ critical region with $n = 144$ is $D_{144} \geqslant 0.8993/\sqrt{144} = 0.0749$.

Finally, the **Kruskal–Wallis** and **Friedman** test statistics are, for large sample sizes, both distributed approximately as the χ^2 distribution with $\nu = k - 1$ degrees of freedom. The appropriate values have been inserted at the ends of the tables on pages 32–34; α_1^R values from the χ^2 table (page 21) are appropriate.

Linear and rank correlation

When data consist of pairs (X, Y) of related measurements it is often important to study whether there is at least an approximate linear relationship between X and Y. The strength of such a relationship is measured by the linear correlation coefficient ρ (rho), which always lies between -1 and $+1$. $\rho = 0$ indicates no linear relationship; $\rho = +1$ and $\rho = -1$ indicate exact linear relationships of $+$ve and $-$ve slopes respectively. More generally, values of ρ near 0 indicate little linear relationship, and values near $+1$ or -1 indicate strong linear relationships.

Tests, etc. concerning ρ are formulated using the sample linear correlation coefficient $r = \Sigma(X - \bar{X})(Y - \bar{Y})/\{\Sigma(X - \bar{X})^2 \Sigma(Y - \bar{Y})^2\}^{1/2}$, $\bar{X}$ and $\bar{Y}$ being the sample mean values of X and Y. The first table on page 36 is for testing the null hypothesis H_0 that $\rho = 0$. Critical regions are $|r| \geqslant$ tabulated value if H_1 is the two-sided alternative hypothesis $\rho \neq 0$ (using significance levels α_2) or $r \geqslant$ tabulated value or $r \leqslant -$(tabulated value) if H_1 is $\rho > 0$ or $\rho < 0$ respectively (using levels α_1^R).

The following data show the market value (in units of £10 000) of eight houses four years ago (X) and currently (Y).

X	0.8	1.7	2.4	0.9	1.2	1.6	1.7	2.9
Y	1.3	3.3	3.8	1.1	2.4	3.1	3.5	3.9

Here r is found to be 0.8918. This is very strong evidence in favour of

the one-sided H_1: $\rho > 0$, since the $\alpha_1^R = \frac{1}{2}\%$ critical region with sample size $n = 8$ is $r \geqslant 0.8343$. Had α_1^L critical values been required, they would have been given by the α_1^R values prefixed with a minus sign.

The construction of confidence intervals for ρ and the testing of values of ρ other than $\rho = 0$ may be accomplished using Fisher's z-transformation. For any value of r or ρ, this gives a 'z-value', $z(r)$ or $z(\rho)$, computed from

$$z(r) = \frac{1}{2}\log_e\left(\frac{1+r}{1-r}\right) = 1.1513 \log_{10}\left(\frac{1+r}{1-r}\right)$$

and $z(r)$ is known to have an approximate normal distribution with mean $z(\rho)$ and standard deviation $1/\sqrt{n-3}$. A table giving $z(r)$ is provided on page 36, and on page 37 there is a table for converting back from a z-value to its corresponding r-value or ρ-value. If r or ρ is $-$ve, attach a minus sign to the z-value, and vice versa.

So to find a $\gamma = 95\%$ confidence interval for ρ with the above data, we first find the 95% confidence interval for $z(\rho)$ as $\{z(r) - 1.9600/\sqrt{n-3} : z(r) + 1.9600/\sqrt{n-3}\}$ (the 1.9600 being the $\gamma = 95\%$ value in the table of normal percentage points on page 20) where $n = 8$ and $z(r) = z(0.8918)$, which is about 1.4306 (interpolating between $z(0.891) = 1.4268$ and $z(0.892) = 1.4316$ on page 36). This interval works out to (0.554 : 2.307). These limits for the value of $z(\rho)$ are then converted to ρ-values by the table on page 37, giving the confidence interval for ρ of (0.503 : 0.980). As a second example, if we wish to test H_0: $\rho = 0.8$ against H_1: $\rho > 0.8$ at the $\alpha_1^R = 5\%$ significance level, the critical value for $z(r)$ would be $z(0.8) + 1.6449/\sqrt{n-3} = 1.0986 + 1.6449/\sqrt{5} = 1.834$ (the 1.6449 again coming from page 20). The critical region $z(r) \geqslant 1.834$ then converts to $r \geqslant 0.950$ from page 37, and so we are unable to reject H_0: $\rho = 0.8$ in favour of H_1: $\rho > 0.8$ at this significance level.

An alternative and quicker method is to use the charts on pages 38–39. For confidence intervals, locate the obtained value of r on the horizontal axis, trace along the vertical to the points of intersection with the two curves labelled with the sample size n, and read off the confidence limits on the vertical axis. For critical values, locate the hypothesised value of ρ, say ρ_0, on the vertical axis, trace along the horizontal to the points of intersection with the two curves, and read off the critical values on the horizontal axis. If these two values are r_1 and r_2, with $r_1 < r_2$, then the one-sided critical regions with significance level α_1 for testing H_0: $\rho = \rho_0$ against H_1: $\rho < \rho_0$ or H_1: $\rho > \rho_0$ are $r \leqslant r_1$ and $r \geqslant r_2$ respectively, and the critical region with significance level $\alpha_2 = 2\alpha_1$ for testing H_0 against H_1: $\rho \neq \rho_0$ is comprised of both of these one-sided regions.

The reader may check the charts for the results found above using the z-transformation. Accuracy may be rather limited, especially when r and ρ are close to $+1$ or -1; however the z-transformation methods are not completely accurate either, especially for small n. Further inaccuracies may occur for sample sizes not included on the charts, in which case the user has to judge distances between the curves.

All of the above work depends on the assumption that (X, Y) has a bivariate normal distribution. Tables for two nonparametric methods, which do not require such an assumption, are given on page 40. These methods do not test specifically for linearity but for the tendency of Y to increase (or decrease) as X increases.

To calculate **Spearman's** rank correlation coefficient, first rank the X-values and Y-values separately from 1 to n, calculate the difference in ranks for each (X, Y) pair, and sum the squares of these differences to obtain D^2. Spearman's coefficient r_S is calculated as $r_S = 1 - 6D^2/(n^3 - n)$. With the above data we have:

X-ranks	1	$5\frac{1}{2}$	7	2	3	4	$5\frac{1}{2}$	8
Y-ranks	2	5	7	1	3	4	6	8
rank differences	-1	$\frac{1}{2}$	0	1	0	0	$-\frac{1}{2}$	0

Thus D^2 is $2(1)^2 + 2(\frac{1}{2})^2 + 4(0)^2 = 2\frac{1}{2}$, giving $r_S = 1 - 6 \times 2\frac{1}{2}/(8^3 - 8) = 0.9702$. The $\alpha_1^R = \frac{1}{2}\%$ critical region for testing against the tendency for Y to increase with X is $r_S \geqslant 0.8810$, so there is virtually conclusive proof that this tendency is present. The general forms of the critical regions are the same as for r above.

For **Kendall's** rank correlation coefficient, we compare each (X, Y) pair in turn with every other pair; if the pair with the smaller X-value also has the smaller Y-value, the pair is said to be *concordant*, but if it has the larger Y-value the pair is *discordant*. If N_C and N_D are the total numbers of concordant and discordant pairs, Kendall's coefficient τ is calculated as $\tau = (N_C - N_D)/\frac{1}{2}n(n-1)$, where in fact $\frac{1}{2}n(n-1)$ is the total number of comparisons made. Any comparison in which the X-values and/or the Y-values are equal counts $\frac{1}{2}$ to both N_C and N_D. Critical regions are of the same forms as with r and r_S. In the above example, $N_C = 26\frac{1}{2}$, $N_D = 1\frac{1}{2}$, and $\tau = (26\frac{1}{2} - 1\frac{1}{2})/28 = 0.8929$. This is again clearly significant of the tendency for Y to increase with X, since the $\alpha_1^R = \frac{1}{2}\%$ critical region is $\tau \geqslant 0.7857$.

Critical regions for large n may be found using the facts that, under the null hypothesis, r, r_S and τ have approximate normal distributions with zero means and standard deviations $1/\sqrt{n-1}$ for both r and r_S, and $\{2(2n + 5)/9n(n-1)\}^{1/2}$ for τ. For example the reader may check that with $n = 144$ the approximate $\alpha_2 = 5\%$ critical regions are $|r| \geqslant 0.1639$, $|r_S| \geqslant 0.1639$ and $|\tau| \geqslant 0.1102$.

Critical values for the sample linear correlation coefficient r

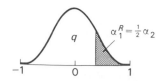

q	0.95	0.975	0.99	0.995
α_1^R	5%	2½%	1%	½%
α_2	10%	5%	2%	1%
n				
1	–	–	–	–
2	–	–	–	–
3	0.9877	0.9969	0.9995	0.9999
4	0.9000	0.9500	0.9800	0.9900
5	0.8054	0.8783	0.9343	0.9587
6	0.7293	0.8114	0.8822	0.9172
7	0.6694	0.7545	0.8329	0.8745
8	0.6215	0.7067	0.7887	0.8343
9	0.5822	0.6664	0.7498	0.7977
10	0.5494	0.6319	0.7155	0.7646
11	0.5214	0.6021	0.6851	0.7348
12	0.4973	0.5760	0.6581	0.7079
13	0.4762	0.5529	0.6339	0.6835
14	0.4575	0.5324	0.6120	0.6614
15	0.4409	0.5140	0.5923	0.6411
16	0.4259	0.4973	0.5742	0.6226
17	0.4124	0.4821	0.5577	0.6055
18	0.4000	0.4683	0.5425	0.5897
19	0.3887	0.4555	0.5285	0.5751
20	0.3783	0.4438	0.5155	0.5614
21	0.3687	0.4329	0.5034	0.5487
22	0.3598	0.4227	0.4921	0.5368
23	0.3515	0.4132	0.4815	0.5256
24	0.3438	0.4044	0.4716	0.5151
25	0.3365	0.3961	0.4622	0.5052
26	0.3297	0.3882	0.4534	0.4958
27	0.3233	0.3809	0.4451	0.4869
28	0.3172	0.3739	0.4372	0.4785
29	0.3115	0.3673	0.4297	0.4705
30	0.3061	0.3610	0.4226	0.4629

q	0.95	0.975	0.99	0.995
α_1^R	5%	2½%	1%	½%
α_2	10%	5%	2%	1%
n				
31	0.3009	0.3550	0.4158	0.4556
32	0.2960	0.3494	0.4093	0.4487
33	0.2913	0.3440	0.4032	0.4421
34	0.2869	0.3388	0.3972	0.4357
35	0.2826	0.3338	0.3916	0.4296
36	0.2785	0.3291	0.3862	0.4238
37	0.2746	0.3246	0.3810	0.4182
38	0.2709	0.3202	0.3760	0.4128
39	0.2673	0.3160	0.3712	0.4076
40	0.2638	0.3120	0.3665	0.4026
41	0.2605	0.3081	0.3621	0.3978
42	0.2573	0.3044	0.3578	0.3932
43	0.2542	0.3008	0.3536	0.3887
44	0.2512	0.2973	0.3496	0.3843
45	0.2483	0.2940	0.3457	0.3801
46	0.2455	0.2907	0.3420	0.3761
47	0.2429	0.2876	0.3384	0.3721
48	0.2403	0.2845	0.3348	0.3683
49	0.2377	0.2816	0.3314	0.3646
50	0.2353	0.2787	0.3281	0.3610
51	0.2329	0.2759	0.3249	0.3575
52	0.2306	0.2732	0.3218	0.3542
53	0.2284	0.2706	0.3188	0.3509
54	0.2262	0.2681	0.3158	0.3477
55	0.2241	0.2656	0.3129	0.3445
56	0.2221	0.2632	0.3102	0.3415
57	0.2201	0.2609	0.3074	0.3385
58	0.2181	0.2586	0.3048	0.3357
59	0.2162	0.2564	0.3022	0.3328
60	0.2144	0.2542	0.2997	0.3301

q	0.95	0.975	0.99	0.995
α_1^R	5%	2½%	1%	½%
α_2	10%	5%	2%	1%
n				
61	0.2126	0.2521	0.2972	0.3274
62	0.2108	0.2500	0.2948	0.3248
63	0.2091	0.2480	0.2925	0.3223
64	0.2075	0.2461	0.2902	0.3198
65	0.2058	0.2441	0.2880	0.3173
66	0.2042	0.2423	0.2858	0.3150
67	0.2027	0.2404	0.2837	0.3126
68	0.2012	0.2387	0.2816	0.3104
69	0.1997	0.2369	0.2796	0.3081
70	0.1982	0.2352	0.2776	0.3060
71	0.1968	0.2335	0.2756	0.3038
72	0.1954	0.2319	0.2737	0.3017
73	0.1940	0.2303	0.2718	0.2997
74	0.1927	0.2287	0.2700	0.2977
75	0.1914	0.2272	0.2682	0.2957
76	0.1901	0.2257	0.2664	0.2938
77	0.1888	0.2242	0.2647	0.2919
78	0.1876	0.2227	0.2630	0.2900
79	0.1864	0.2213	0.2613	0.2882
80	0.1852	0.2199	0.2597	0.2864
82	0.1829	0.2172	0.2565	0.2830
84	0.1807	0.2146	0.2535	0.2796
86	0.1786	0.2120	0.2505	0.2764
88	0.1765	0.2096	0.2477	0.2732
90	0.1745	0.2072	0.2449	0.2702
92	0.1726	0.2050	0.2422	0.2673
94	0.1707	0.2028	0.2396	0.2645
96	0.1689	0.2006	0.2371	0.2617
98	0.1671	0.1986	0.2347	0.2591
100	0.1654	0.1966	0.2324	0.2565

For description, see page 35.

The Fisher z-transformation

$$z(r) = \tfrac{1}{2} \log_e\left(\frac{1+r}{1-r}\right) = 1.1513 \log_{10}\left(\frac{1+r}{1-r}\right)$$

r	0	1	2	3	4	5	6	7	8	9
0.0	0.0000	0.0100	0.0200	0.0300	0.0400	0.0500	0.0601	0.0701	0.0802	0.0902
0.1	0.1003	0.1104	0.1206	0.1307	0.1409	0.1511	0.1614	0.1717	0.1820	0.1923
0.2	0.2027	0.2132	0.2237	0.2342	0.2448	0.2554	0.2661	0.2769	0.2877	0.2986
0.3	0.3095	0.3205	0.3316	0.3428	0.3541	0.3654	0.3769	0.3884	0.4001	0.4118
0.4	0.4236	0.4356	0.4477	0.4599	0.4722	0.4847	0.4973	0.5101	0.5230	0.5361
0.5	0.5493	0.5627	0.5763	0.5901	0.6042	0.6184	0.6328	0.6475	0.6625	0.6777
0.6	0.6931	0.7089	0.7250	0.7414	0.7582	0.7753	0.7928	0.8107	0.8291	0.8480
0.7	0.8673	0.8872	0.9076	0.9287	0.9505	0.9730	0.9962	1.0203	1.0454	1.0714
0.80	1.0986	1.1014	1.1042	1.1070	1.1098	1.1127	1.1155	1.1184	1.1212	1.1241
0.81	1.1270	1.1299	1.1329	1.1358	1.1388	1.1417	1.1447	1.1477	1.1507	1.1538
0.82	1.1568	1.1599	1.1630	1.1660	1.1692	1.1723	1.1754	1.1786	1.1817	1.1849
0.83	1.1881	1.1914	1.1946	1.1979	1.2011	1.2044	1.2077	1.2111	1.2144	1.2178
0.84	1.2212	1.2246	1.2280	1.2315	1.2349	1.2384	1.2419	1.2454	1.2490	1.2526
0.85	1.2562	1.2598	1.2634	1.2671	1.2707	1.2745	1.2782	1.2819	1.2857	1.2895
0.86	1.2933	1.2972	1.3011	1.3050	1.3089	1.3129	1.3169	1.3209	1.3249	1.3290
0.87	1.3331	1.3372	1.3414	1.3456	1.3498	1.3540	1.3583	1.3626	1.3670	1.3714
0.88	1.3758	1.3802	1.3847	1.3892	1.3938	1.3984	1.4030	1.4077	1.4124	1.4171
0.89	1.4219	1.4268	1.4316	1.4365	1.4415	1.4465	1.4516	1.4566	1.4618	1.4670
0.90	1.4722	1.4775	1.4828	1.4882	1.4937	1.4992	1.5047	1.5103	1.5160	1.5217
0.91	1.5275	1.5334	1.5393	1.5453	1.5513	1.5574	1.5636	1.5698	1.5762	1.5826
0.92	1.5890	1.5956	1.6022	1.6089	1.6157	1.6226	1.6296	1.6366	1.6438	1.6510
0.93	1.6584	1.6658	1.6734	1.6811	1.6888	1.6967	1.7047	1.7129	1.7211	1.7295
0.94	1.7380	1.7467	1.7555	1.7645	1.7736	1.7828	1.7923	1.8019	1.8117	1.8216
0.95	1.8318	1.8421	1.8527	1.8635	1.8745	1.8857	1.8972	1.9090	1.9210	1.9333
0.96	1.9459	1.9588	1.9721	1.9857	1.9996	2.0139	2.0287	2.0439	2.0595	2.0756
0.97	2.0923	2.1095	2.1273	2.1457	2.1649	2.1847	2.2054	2.2269	2.2494	2.2729
0.98	2.2976	2.3235	2.3507	2.3796	2.4101	2.4427	2.4774	2.5147	2.5550	2.5987
0.99	2.6467	2.6996	2.7587	2.8257	2.9031	2.9945	3.1063	3.2504	3.4534	3.8002

For description, see page 35.

The inverse of the Fisher z-transformation

z	0	1	2	3	4	5	6	7	8	9
0.0	0.0000	0100	0200	0300	0400	0500	0599	0699	0798	0898
0.1	0.0997	1096	1194	1293	1391	1489	1586	1684	1781	1877
0.2	0.1974	2070	2165	2260	2355	2449	2543	2636	2729	2821
0.3	0.2913	3004	3095	3185	3275	3364	3452	3540	3627	3714
0.4	0.3799	3885	3969	4053	4136	4219	4301	4382	4462	4542
0.5	0.4621	4699	4777	4854	4930	5005	5080	5154	5227	5299
0.6	0.5370	5441	5511	5581	5649	5717	5784	5850	5915	5980
0.7	0.6044	6107	6169	6231	6291	6351	6411	6469	6527	6584
0.8	0.6640	6696	6751	6805	6858	6911	6963	7014	7064	7114
0.9	0.7163	7211	7259	7306	7352	7398	7443	7487	7531	7574
1.0	0.7616	7658	7699	7739	7779	7818	7857	7895	7932	7969
1.1	0.8005	8041	8076	8110	8144	8178	8210	8243	8275	8306
1.2	0.8337	8367	8397	8426	8455	8483	8511	8538	8565	8591
1.3	0.8617	8643	8668	8692	8717	8741	8764	8787	8810	8832
1.4	0.8854	8875	8896	8917	8937	8957	8977	8996	9016	9033
1.5	0.9051	9069	9087	9104	9121	9138	9154	9170	9186	9201
1.6	0.9217	9232	9246	9261	9275	9289	9302	9316	9329	9341
1.7	0.9354	9366	9379	9391	9402	9414	9425	9436	9447	9458
1.8	0.9468	9478	9488	9498	9508	9517	9527	9536	9545	9554
1.9	0.9562	9571	9579	9587	9595	9603	9611	9618	9626	9633
2.0	0.9640	9647	9654	9661	9667	9674	9680	9687	9693	9699
2.1	0.9705	9710	9716	9721	9727	9732	9737	9743	9748	9753
2.2	0.9757	9762	9767	9771	9776	9780	9785	9789	9793	9797
2.3	0.9801	9805	9809	9812	9816	9820	9823	9827	9830	9833
2.4	0.9837	9840	9843	9846	9849	9852	9855	9858	9861	9863
2.5	0.9866	9869	9871	9874	9876	9879	9881	9884	9886	9888
2.6	0.9890	9892	9895	9897	9899	9901	9903	9905	9906	9908
2.7	0.9910	9912	9914	9915	9917	9919	9920	9922	9923	9925
2.8	0.9926	9928	9929	9931	9932	9933	9935	9936	9937	9938
2.9	0.9940	9941	9942	9943	9944	9945	9946	9947	9949	9950

ADD PROPORTIONAL PARTS

The proportional parts appear in two interleaved lines per z interval (the first aligned with the z value row, the second aligned with the intermediate 5–9 line).

ref	1	2	3	4	5	6	7	8	9
0.0	10	20	30	40	50	60	70	80	90
0500	10	20	30	40	50	60	70	80	89
0.1	10	20	30	39	49	59	69	79	89
1489	10	19	29	39	48	58	68	78	87
0.2	10	19	28	38	48	57	66	76	86
2449	9	19	28	37	46	56	65	74	84
0.3	9	18	27	36	45	54	63	72	81
3364	9	17	26	35	44	52	61	70	78
0.4	8	17	25	34	42	50	59	67	76
4219	8	16	24	32	40	48	56	64	72
0.5	8	15	23	31	38	46	54	61	69
5005	7	15	22	29	36	44	51	58	66
0.6	7	14	21	28	35	42	49	56	62
5717	7	13	20	26	33	39	46	52	59
0.7	6	12	18	25	31	37	43	49	55
6351	6	12	17	23	29	35	40	46	52
0.8	5	11	16	22	27	33	38	43	49
6911	5	10	15	20	25	30	35	40	45
0.9	5	9	14	19	24	28	33	38	42
7398	4	9	13	17	22	26	31	35	39
1.0	4	8	12	16	20	24	28	32	36
7818	4	7	11	15	19	22	26	30	34
1.1	3	7	10	14	17	21	24	28	31
8178	3	6	10	13	16	19	22	25	29
1.2	3	6	9	12	15	18	20	23	26
8483	3	5	8	11	13	16	19	21	24
1.3	2	5	7	10	12	15	17	20	22
8741	2	5	7	9	11	14	16	18	20
1.4	2	4	6	8	10	12	14	16	19
8957	2	4	6	8	9	11	13	15	17
1.5	2	3	5	7	9	10	12	14	16
9138	2	3	5	6	8	9	11	13	14
1.6	1	3	4	6	7	9	10	12	13
9289	1	3	4	5	6	8	9	10	12
1.7	1	2	4	5	6	7	8	10	11
9414	1	2	3	4	5	6	8	9	10
1.8	1	2	3	4	5	6	7	8	9
9517	1	2	3	4	4	5	6	7	8
1.9	1	2	2	3	4	5	6	7	7
9603	1	1	2	3	4	4	5	6	7
2.0	1	1	2	3	3	4	5	5	6
9674	1	1	2	2	3	4	4	5	6
2.1	1	1	2	2	3	3	4	4	5
9732	0	1	2	2	2	3	4	4	4
2.2	0	1	1	2	2	3	3	4	4
9780	0	1	1	2	2	3	3	3	4
2.3	0	1	1	2	2	2	3	3	3
9820	0	1	1	1	2	2	2	3	3
2.4	0	1	1	1	2	2	2	2	3
9852	0	1	1	1	1	2	2	2	3
2.5	0	0	1	1	1	1	2	2	2
2.6	0	0	1	1	1	1	1	2	2
2.7	0	0	0	1	1	1	1	1	1
2.8	0	0	0	1	1	1	1	1	1
2.9	0	0	0	0	1	1	1	1	1

z	0	1	2	3	4	5	6	7	8	9
3.	0.995055	0.995949	0.996682	0.997283	0.997775	0.998178	0.998508	0.998778	0.999000	0.999181
4.	0.999329	0.999451	0.999550	0.999632	0.999699	0.999753	0.999798	0.999835	0.999865	0.999889
5.	0.999909	0.999926	0.999939	0.999950	0.999959	0.999967	0.999973	0.999978	0.999982	0.999985
6.	0.999988	0.999990	0.999992	0.999993	0.999994	0.999995	0.999996	0.999997	0.999998	0.999998
7.	0.999998	0.999999	0.999999	0.999999	0.999999	0.999999	0.999999	1.000000	1.000000	1.000000

For description, see page 35.

Charts giving confidence intervals for ρ and critical values for r

| $\alpha_1 = 5\%$ | $\alpha_2 = 10\%$ | $\gamma = 90\%$ |

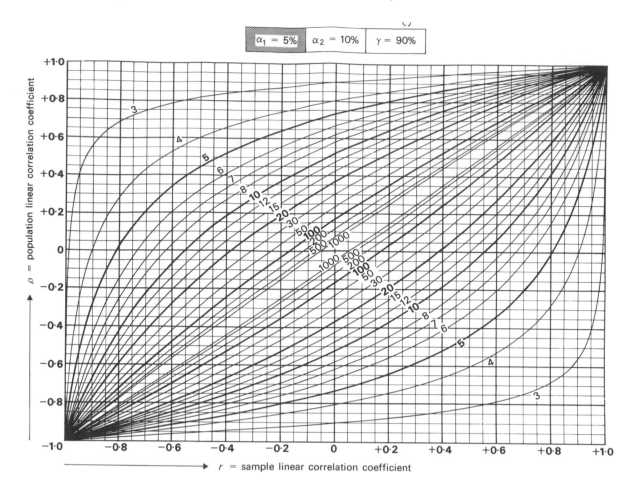

r = sample linear correlation coefficient

| $\alpha_1 = 2\frac{1}{2}\%$ | $\alpha_2 = 5\%$ | $\gamma = 95\%$ |

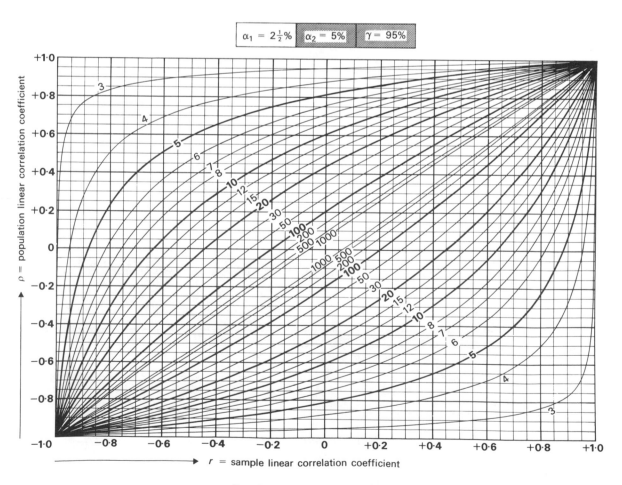

r = sample linear correlation coefficient

For description, see page 35.

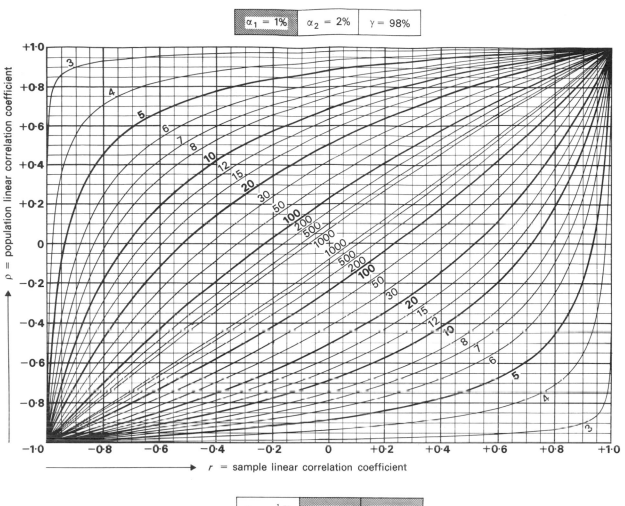

| $\alpha_1 = 1\%$ | $\alpha_2 = 2\%$ | $\gamma = 98\%$ |

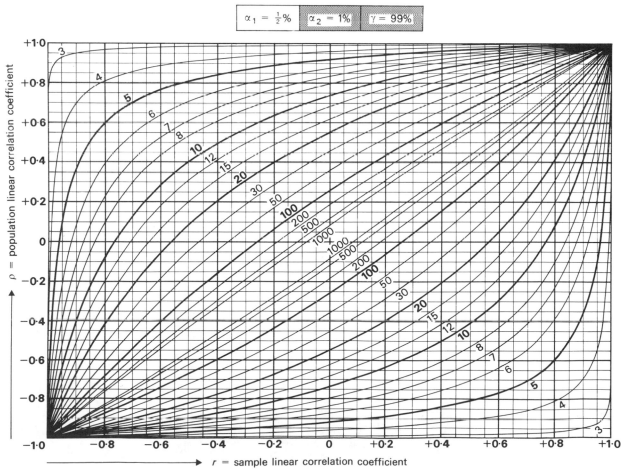

| $\alpha_1 = \frac{1}{2}\%$ | $\alpha_2 = 1\%$ | $\gamma = 99\%$ |

For description, see page 35.

Critical values for Spearman's rank correlation coefficient

$$r_S = 1 - \frac{6D^2}{n^3 - n}$$

α_1^R	5%	2½%	1%	½%
α_2	10%	5%	2%	1%
n				
1	–	–	–	–
2	–	–	–	–
3	–	–	–	–
4	1.0000	–	–	–
5	0.9000	1.0000	1.0000	–
6	0.8286	0.8857	0.9429	1.0000
7	0.7143	0.7857	0.8929	0.9286
8	0.6429	0.7381	0.8333	0.8810
9	0.6000	0.7000	0.7833	0.8333
10	0.5636	0.6485	0.7455	0.7939
11	0.5364	0.6182	0.7091	0.7545
12	0.5035	0.5874	0.6783	0.7273
13	0.4835	0.5604	0.6484	0.7033
14	0.4637	0.5385	0.6264	0.6791
15	0.4464	0.5214	0.6036	0.6536
16	0.4294	0.5029	0.5824	0.6353
17	0.4142	0.4877	0.5662	0.6176
18	0.4014	0.4716	0.5501	0.5996
19	0.3912	0.4596	0.5351	0.5842
20	0.3805	0.4466	0.5218	0.5699
21	0.3701	0.4364	0.5091	0.5558
22	0.3608	0.4252	0.4975	0.5438
23	0.3528	0.4160	0.4862	0.5316
24	0.3443	0.4070	0.4757	0.5209
25	0.3369	0.3977	0.4662	0.5108
26	0.3306	0.3901	0.4571	0.5009
27	0.3242	0.3828	0.4487	0.4915
28	0.3180	0.3755	0.4401	0.4828
29	0.3118	0.3685	0.4325	0.4749
30	0.3063	0.3624	0.4251	0.4670
31	0.3012	0.3560	0.4185	0.4593
32	0.2962	0.3504	0.4117	0.4523
33	0.2914	0.3449	0.4054	0.4455
34	0.2871	0.3396	0.3995	0.4390
35	0.2829	0.3347	0.3936	0.4328
36	0.2788	0.3300	0.3882	0.4268
37	0.2748	0.3253	0.3829	0.4211
38	0.2710	0.3209	0.3778	0.4155
39	0.2674	0.3168	0.3729	0.4103
40	0.2640	0.3128	0.3681	0.4051
41	0.2606	0.3087	0.3636	0.4002
42	0.2574	0.3051	0.3594	0.3955
43	0.2543	0.3014	0.3550	0.3908
44	0.2513	0.2978	0.3511	0.3865
45	0.2484	0.2945	0.3470	0.3822
46	0.2456	0.2913	0.3433	0.3781
47	0.2429	0.2880	0.3396	0.3741
48	0.2403	0.2850	0.3361	0.3702
49	0.2378	0.2820	0.3326	0.3664
50	0.2353	0.2791	0.3293	0.3628
51	0.2329	0.2764	0.3260	0.3592
52	0.2307	0.2736	0.3228	0.3558
53	0.2284	0.2710	0.3198	0.3524
54	0.2262	0.2685	0.3168	0.3492
55	0.2242	0.2659	0.3139	0.3460
56	0.2221	0.2636	0.3111	0.3429
57	0.2201	0.2612	0.3083	0.3400
58	0.2181	0.2589	0.3057	0.3370
59	0.2162	0.2567	0.3030	0.3342
60	0.2144	0.2545	0.3005	0.3314
61	0.2126	0.2524	0.2980	0.3287
62	0.2108	0.2503	0.2956	0.3260
63	0.2091	0.2483	0.2933	0.3234
64	0.2075	0.2463	0.2910	0.3209
65	0.2058	0.2444	0.2887	0.3185
66	0.2042	0.2425	0.2865	0.3161
67	0.2027	0.2407	0.2844	0.3137
68	0.2012	0.2389	0.2823	0.3114
69	0.1997	0.2372	0.2802	0.3092
70	0.1982	0.2354	0.2782	0.3070
71	0.1968	0.2337	0.2762	0.3048
72	0.1954	0.2321	0.2743	0.3027
73	0.1940	0.2305	0.2724	0.3006
74	0.1927	0.2289	0.2706	0.2986
75	0.1914	0.2274	0.2688	0.2966
76	0.1901	0.2259	0.2670	0.2947
77	0.1888	0.2244	0.2652	0.2928
78	0.1876	0.2229	0.2635	0.2909
79	0.1864	0.2215	0.2619	0.2891
80	0.1852	0.2201	0.2602	0.2872
82	0.1829	0.2174	0.2570	0.2837
84	0.1807	0.2147	0.2539	0.2804
86	0.1785	0.2122	0.2510	0.2771
88	0.1765	0.2097	0.2481	0.2740
90	0.1745	0.2074	0.2453	0.2709
92	0.1725	0.2051	0.2426	0.2680
94	0.1707	0.2029	0.2400	0.2651
96	0.1689	0.2008	0.2375	0.2623
98	0.1671	0.1987	0.2351	0.2597
100	0.1654	0.1967	0.2327	0.2571

For description, see page 35.

Critical values for Kendall's rank correlation coefficient

$$\tau = \frac{N_C - N_D}{\frac{1}{2}n(n-1)}$$

α_1^R	5%	2½%	1%	½%
α_2	10%	5%	2%	1%
n				
1	–	–	–	–
2	–	–	–	–
3	–	–	–	–
4	1.0000	–	–	–
5	0.8000	1.0000	1.0000	–
6	0.7333	0.8667	0.8667	1.0000
7	0.6190	0.7143	0.8095	0.9048
8	0.5714	0.6429	0.7143	0.7857
9	0.5000	0.5556	0.6667	0.7222
10	0.4667	0.5111	0.6000	0.6444
11	0.4182	0.4909	0.5636	0.6000
12	0.3939	0.4545	0.5455	0.5758
13	0.3590	0.4359	0.5128	0.5641
14	0.3626	0.4066	0.4725	0.5165
15	0.3333	0.3905	0.4667	0.5048
16	0.3167	0.3833	0.4333	0.4833
17	0.3088	0.3676	0.4265	0.4706
18	0.2941	0.3464	0.4118	0.4510
19	0.2865	0.3333	0.3918	0.4386
20	0.2737	0.3263	0.3789	0.4211
21	0.2667	0.3143	0.3714	0.4095
22	0.2641	0.3074	0.3593	0.3939
23	0.2569	0.2964	0.3518	0.3913
24	0.2464	0.2899	0.3406	0.3768
25	0.2400	0.2867	0.3333	0.3667
26	0.2369	0.2800	0.3292	0.3600
27	0.2308	0.2707	0.3219	0.3561
28	0.2275	0.2646	0.3122	0.3439
29	0.2217	0.2611	0.3103	0.3399
30	0.2184	0.2552	0.3011	0.3333
31	0.2129	0.2516	0.2946	0.3247
32	0.2097	0.2460	0.2903	0.3226
33	0.2045	0.2424	0.2879	0.3144
34	0.2014	0.2371	0.2799	0.3119
35	0.1966	0.2336	0.2773	0.3042
36	0.1937	0.2317	0.2730	0.3016
37	0.1922	0.2282	0.2673	0.2973
38	0.1892	0.2233	0.2632	0.2916
39	0.1876	0.2200	0.2605	0.2874
40	0.1846	0.2179	0.2564	0.2846
41	0.1805	0.2146	0.2537	0.2805
42	0.1777	0.2125	0.2497	0.2753
43	0.1761	0.2093	0.2470	0.2735
44	0.1734	0.2072	0.2431	0.2685
45	0.1717	0.2040	0.2404	0.2667
46	0.1691	0.2019	0.2386	0.2638
47	0.1674	0.1989	0.2359	0.2599
48	0.1667	0.1968	0.2323	0.2571
49	0.1633	0.1956	0.2296	0.2534
50	0.1624	0.1918	0.2278	0.2506
51	0.1608	0.1906	0.2251	0.2486
52	0.1584	0.1885	0.2232	0.2459
53	0.1567	0.1872	0.2206	0.2438
54	0.1558	0.1852	0.2187	0.2411
55	0.1542	0.1825	0.2162	0.2391
56	0.1519	0.1805	0.2143	0.2364
57	0.1516	0.1792	0.2118	0.2343
58	0.1494	0.1773	0.2099	0.2317
59	0.1479	0.1759	0.2086	0.2297
60	0.1469	0.1740	0.2068	0.2282
61	0.1454	0.1727	0.2044	0.2262
62	0.1444	0.1719	0.2025	0.2237
63	0.1429	0.1705	0.2012	0.2227
64	0.1419	0.1687	0.1994	0.2202
65	0.1404	0.1673	0.1981	0.2183
66	0.1394	0.1655	0.1963	0.2168
67	0.1389	0.1642	0.1949	0.2148
68	0.1370	0.1633	0.1932	0.2133
69	0.1364	0.1620	0.1918	0.2114
70	0.1354	0.1611	0.1901	0.2099
71	0.1340	0.1598	0.1887	0.2089
72	0.1330	0.1581	0.1878	0.2074
73	0.1324	0.1575	0.1865	0.2055
74	0.1314	0.1559	0.1847	0.2040
75	0.1301	0.1553	0.1834	0.2029
76	0.1291	0.1537	0.1825	0.2014
77	0.1285	0.1531	0.1811	0.2003
78	0.1275	0.1515	0.1795	0.1988
79	0.1269	0.1509	0.1788	0.1970
80	0.1259	0.1500	0.1772	0.1962
82	0.1244	0.1478	0.1749	0.1936
84	0.1228	0.1457	0.1727	0.1910
86	0.1212	0.1442	0.1710	0.1885
88	0.1196	0.1426	0.1688	0.1865
90	0.1186	0.1406	0.1665	0.1845
92	0.1171	0.1390	0.1648	0.1820
94	0.1155	0.1375	0.1631	0.1801
96	0.1145	0.1360	0.1614	0.1785
98	0.1134	0.1349	0.1597	0.1765
100	0.1119	0.1333	0.1580	0.1745

For description, see page 35.

Control chart constants and conversion factors for estimating σ

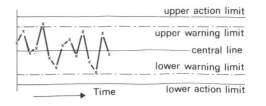

n	W	A	w_1	w_2	a_1	a_2	d_1	d_2	d_3	c
2	1.2282	1.9365	0.0393	2.8092	0.0016	4.1241	1.1284	2.0000	1.4142	0.8862
3	0.6686	1.0541	0.1791	2.1756	0.0356	2.9916	1.6926	2.3391	1.9099	0.5908
4	0.4760	0.7505	0.2888	1.9352	0.0969	2.5787	2.0588	2.5803	2.2346	0.4857
5	0.3768	0.5942	0.3653	1.8045	0.1580	2.3577	2.3259	2.7665	2.4744	0.4299
6	0.3157	0.4978	0.4206	1.7207	0.2110	2.2172	2.5344	2.9177	2.6635	0.3946
7	0.2739	0.4319	0.4624	1.6616	0.2556	2.1187	2.7044	3.0448	2.8189	0.3698
8	0.2434	0.3837	0.4952	1.6173	0.2932	2.0451	2.8472	3.1541	2.9504	0.3512
9	0.2200	0.3468	0.5218	1.5826	0.3251	1.9875	2.9700	3.2499	3.0641	0.3367
10	0.2014	0.3175	0.5438	1.5545	0.3524	1.9410	3.0775	3.3352	3.1640	0.3249
11	0.1863	0.2937	0.5624	1.5312	0.3761	1.9024	3.1729	3.4118	3.2531	0.3152
12	0.1736	0.2738	0.5783	1.5115	0.3969	1.8697	3.2585	3.4815	3.3333	0.3069
13	0.1629	0.2569	0.5922	1.4945	0.4152	1.8417	3.3360	3.5452	3.4061	0.2998
14	0.1538	0.2424	0.6044	1.4796	0.4316	1.8172	3.4068	3.6039	3.4728	0.2935
15	0.1458	0.2298	0.6153	1.4666	0.4463	1.7957	3.4718	3.6584	3.5343	0.2880
16	0.1387	0.2187	0.6250	1.4550	0.4596	1.7765	3.5320	3.7091	3.5913	0.2831
17	0.1325	0.2089	0.6338	1.4445	0.4717	1.7592	3.5879	3.7565	3.6443	0.2787
18	0.1269	0.2001	0.6417	1.4351	0.4827	1.7437	3.6401	3.8011	3.6940	0.2747
19	0.1219	0.1922	0.6490	1.4265	0.4928	1.7295	3.6890	3.8430	3.7405	0.2711
20	0.1173	0.1850	0.6557	1.4186	0.5022	1.7165	3.7350	3.8827	3.7844	0.2677

Control charts are designed to aid the regular periodic checking of production and other processes. The situation envisaged is that a quite small sample (the table caters for sample sizes n up to 20) is drawn and examined at regular intervals, and in particular the sample mean $\bar{X}$ and the sample range R are recorded (the *range* is the largest value in the sample minus the smallest value). $\bar{X}$ and R are then plotted on separate control charts to monitor respectively the process average and variability.

The general form of a control chart is illustrated in the diagram. There is a *central line* representing the expected (i.e. average) value of the quantity ($\bar{X}$ or R) being plotted when the process is behaving normally (is *in control*). On either side of the central line are *warning limits* and *action limits*. These terms are virtually self-explanatory. The levels are such that if an observation falls outside the warning limits the user should be alerted to watch the subsequent behaviour of the process but should also realise that such observations are bound to occur by chance occasionally even when the process is in control. An observation may also fall outside the action limits when the process is in control, but the probability of this is very small and so a more positive alert would normally be signalled. Information can also be obtained by watching for possible trends and other such features on the charts.

The central line and warning and action limits may be derived from studying pilot samples taken when the process is presumed or known to be in control, or alternatively may be fixed by *a priori* considerations. If they are derived from pilot samples we shall assume that they are of the same size as those to be taken when the control scheme is in operation and that the mean $\bar{X}$ and range R are calculated for each such sample. These quantities are then averaged over all the pilot samples to obtain $\bar{\bar{X}}$ and $\bar{R}$. We may also calculate, instead of R, either the unadjusted or the adjusted sample standard deviations S or s (see below). The charts are then drawn up as follows:

$\bar{X}$-chart	Central line is $\bar{\bar{X}}$; lower warning limit is $\bar{\bar{X}}-W\bar{R}$; upper warning limit is $\bar{\bar{X}} + W\bar{R}$; lower action limit is $\bar{\bar{X}} - A\bar{R}$; upper action limit is $\bar{\bar{X}} + A\bar{R}$.
R-chart	Central line is $\bar{R}$; lower warning limit is $w_1\bar{R}$; upper warning limit is $w_2\bar{R}$; lower action limit is $a_1\bar{R}$; upper action limit is $a_2\bar{R}$.

As an alternative to using pilot samples, specifications of the mean μ and/or the standard deviation σ of the process measurements may be used to define the 'in control' situation. If μ is given, use it in place of $\bar{\bar{X}}$ in drawing up the $\bar{X}$-chart. If σ is given, the expected value of R is equal to $d_1\sigma$, so here *define* $\bar{R}$ as $d_1\sigma$ and then proceed as above. This application allows an exact interpretation to be made of the warning and action limits, for if the process measurements are normally distributed with mean μ and standard deviation σ the warning limits thus obtained correspond

to quantiles q of 0.025 and 0.975 and the action limits to quantiles of 0.001 and 0.999. In other words, the limits can be regarded as critical values for testing the null hypothesis that the data are indeed from a normal distribution with mean μ and standard deviation σ, the warning limits corresponding to significance levels of $\alpha_1 = 2\frac{1}{2}\%$ or $\alpha_2 = 5\%$ and the action limits to levels of $\alpha_1 = 0.1\%$ or $\alpha_1 = 0.2\%$.

If pilot samples are used it may be that the variability of the process has been measured by recording the sample standard deviations rather than ranges. If the unadjusted sample standard deviation $S = \{\Sigma(X - \bar{X})^2/n\}^{1/2}$ has been calculated for each pilot sample, average the values of S to obtain $\bar{S}$, and then define $\bar{R} = d_2\bar{S}$ and proceed as above. Or, if adjusted sample standard deviations $s = \{\Sigma(X - \bar{X})^2/(n-1)\}^{1/2}$ have been calculated, multiply their average $\bar{s}$ by d_3 to obtain $\bar{R} = d_3\bar{s}$, and again proceed as above. It should be understood that in general these formulae for $\bar{R}$ will not give exactly the same value as if $\bar{R}$ were calculated directly from the pilot samples, but represent the expected value of $\bar{R}$ given the information available.

For convenience we have also included in this table a column of constants c for forming unbiased estimators of the standard deviation σ from either the range of a single sample or the average range of more than one sample of the same size. Denoting by $\bar{R}$ the range or average range, σ is estimated by $c\bar{R}$. σ may also be estimated from $\bar{S}$ or $\bar{s}$ by $cd_2\bar{S}$ or $cd_3\bar{s}$ respectively.

EXAMPLES: If samples are of size $n = 10$, and pilot samples have average value of the sample means $\bar{\bar{X}} = 15.00$ and average range $\bar{R} = 7.00$, then the $\bar{X}$-chart has central line at 15.00, warning limits at $15.00 \pm 0.2014 \times 7.00$, i.e. 13.59 and 16.41, and action limits at $15.00 \pm 0.3175 \times 7.00$, i.e. 12.78 and 17.22; the R-chart has central line at 7.00, warning limits at $0.5438 \times 7.00 = 3.81$ and $1.5545 \times 7.00 = 10.88$, and action limits at $0.3524 \times 7.00 = 2.47$ and $1.9410 \times 7.00 = 13.59$. The standard deviation σ may be estimated from the pilot samples as $c\bar{R} = 0.3249 \times 7.00 = 2.27$.

Alternatively, if the unadjusted sample standard deviations S had been computed instead of ranges, and the average value $\bar{S}$ of the S-values were $\bar{S} = 2.00$, we would *define* $\bar{R} = d_2\bar{S} = 3.3352 \times 2.00 = 6.670$. The reader may confirm that the $\bar{X}$-chart would then have central line 15.00, warning limits 13.66 and 16.34, and action limits 12.88 and 17.12; and the R-chart would have central line 6.670, warning limits 3.63 and 10.37, and action limits 2.35 and 12.95. The standard deviation σ could be estimated as $cd_2\bar{S} = 0.3249 \times 6.670 = 2.17$.

Finally if the 'in control' situation is defined by a mean value $\mu = 14.0$ and standard deviation $\sigma = 2.5$, we *define* $\bar{R} = d_1\sigma = 3.0775 \times 2.5 = 7.694$, and then obtain an $\bar{X}$-chart with central line 14.0, warning limits 12.45 and 15.55, and action limits 11.56 and 16.44; and the R-chart would have central line 7.694, warning limits 4.18 and 11.96, and action limits 2.71 and 14.93.

41

02484	88139	31788	35873	63259	99886	20644	41853	41915	02944	82414	59559	41440	22668	37841	70679	62723	50128	30374	90243
83680	56131	12238	68291	95093	07362	74354	13071	77901	63058	19200	66512	25179	25254	65582	09074	66260	76215	79590	45927
37336	63266	18632	79781	09184	83909	77232	57571	25413	82680	06125	38600	70556	95945	61968	20673	73403	71431	05563	28155
04060	46030	23751	61880	40119	88098	75956	85250	05015	99184	82611	23886	16940	24878	51235	37651	76444	45211	98681	33905
62040	01812	46847	79352	42478	71784	65864	84904	48901	17115	85297	33517	26576	23195	12091	45048	01265	90873	55762	74771
96417	63336	88491	73259	21086	51932	32304	45021	61697	73953	89168	81340	50382	30286	84550	59488	95424	31734	02673	45586
42293	29755	24119	62125	33717	20284	55606	33308	51007	68272	39426	52113	93433	45546	68180	72212	84593	85572	80863	65594
31378	35714	00941	53042	99174	30596	67769	59343	53193	19203	31228	18442	47214	53414	97924	05540	64402	86719	57304	53443
27098	38959	49721	69341	40475	55998	87510	55523	15549	32402	76523	03405	77137	70253	31107	24658	98796	18445	02089	56076
66527	73898	66912	76300	52782	29356	35332	52387	29194	21591	25159	42707	57089	69043	32052	69578	16270	89165	77408	90560
61621	52967	40644	91293	80576	67485	88715	45293	59454	76218	78176	87146	99734	32799	45627	75063	53661	34527	92601	26837
18798	99633	32948	49802	40261	35555	76229	00486	64236	74782	91613	53259	63858	50229	04979	79377	65502	43457	49356	88489
36864	66460	87303	13788	04806	31140	75253	79692	47618	20024	16022	27081	00058	97199	68594	35853	17062	89925	25742	27742
10346	28822	51891	04097	98009	58042	67833	23539	37668	16324	97243	03199	45435	45355	24374	84490	83041	03381	74618	90176
20582	49576	91822	63807	99450	18240	70002	75386	26035	21459	74543	48514	68504	04476	80747	64071	03321	29629	37709	73893
12023	82328	54810	64766	58954	76201	78456	98467	34166	84186	99960	67514	19200	38021	83572	98676	74079	20282	48402	57304
48255	20815	51322	04936	33413	43128	21643	90674	98858	26060	64465	63266	27453	91770	99793	25895	98769	42883	10806	69144
92956	09401	58892	59686	10899	89780	57080	82799	70178	40399	99188	85861	39263	43477	91282	97590	60951	25330	46710	53871
87300	04729	57966	95672	49036	24993	69827	67637	09472	63356	20395	87626	53356	17031	11869	27637	31443	37335	92804	95214
69101	21192	00256	81645	48500	73237	95420	98974	36036	21781	51966	12077	46259	07825	94235	34793	57776	68352	54531	50358
22084	03117	96937	86176	80102	48211	61149	71246	19993	79708	85745	81363	20818	36767	97847	82547	26236	85668	77300	66986
28000	44301	40028	88132	07083	50818	09104	92449	27860	90196	75101	64719	04737	88683	61418	01696	07840	48192	27263	55309
41662	20930	32856	91566	64917	18709	79884	44742	18010	11599	97335	58399	33462	96811	90330	45280	21168	10926	79370	17080
91398	16841	51399	82654	00857	21068	94121	39197	27752	67308	12973	35509	97578	08528	07939	19501	39093	82060	36813	43665
46560	00597	84561	42334	06695	26306	16832	63140	13762	15598	91443	82220	90671	08547	56540	08344	54851	89257	50154	35280
61673	61959	54745	84399	22441	71993	24053	40677	75150	51292	09945	74989	81255	84439	45915	13741	77501	73833	13243	13690
25278	30989	97503	74974	17877	35496	58987	29194	25288	83687	62479	21613	23712	95585	61708	17373	69578	61261	83411	50916
82490	86291	27290	55596	95034	40588	63015	06872	56579	25469	85728	37237	40103	50433	59150	84496	42377	53768	52138	66811
85582	09721	32438	88402	69377	39643	42119	18649	83509	85186	14785	04821	19119	08325	26905	24560	08833	55675	76433	08759
28396	63296	73130	18400	02901	82926	78554	90463	25440	81318	71414	27979	12176	01123	07778	47874	32190	74583	17331	67238
59998	50022	98409	54261	50134	26029	67725	44121	23525	88968	57474	72693	31564	16376	54678	62080	22427	41598	17111	51552
22492	73949	98852	55637	60230	33538	48182	97752	47814	90825	23256	62925	93618	97895	78380	86844	93722	61372	27910	00659
39349	35856	67457	31696	26702	11732	45207	69441	62834	32190	01596	93921	72858	01296	55194	79853	26796	37966	78296	84728
65672	89737	99330	52248	12804	45281	62560	64392	54661	75644	90682	05676	62846	91320	04486	12001	51814	25320	77870	13884
22482	41532	17809	99677	77013	22795	79476	68805	01511	14238	28255	67944	48065	84609	64977	54467	02416	21482	09980	61422
45307	95424	17664	96768	01289	13413	37732	46527	65156	33008	74354	17027	00244	26018	91737	95362	61323	63663	96118	79751
04771	44497	61709	82465	56798	01632	83576	87547	13795	07104	05742	03616	94098	38561	09721	38603	44622	86735	29208	88356
75612	61553	02595	24676	49317	00084	58196	40422	30294	90874	69516	10014	13424	06670	91354	02759	27300	80870	13923	94134
62321	72533	23418	06305	41547	40150	55300	23898	34891	85908	43049	24210	60559	39576	21993	82807	18533	24684	16992	15688
06318	89138	46129	47950	73947	87945	81956	06171	30239	77245	49200	67528	12476	16409	74959	12940	23735	78506	66269	74396
23176	39056	42285	87925	71241	48538	16124	13541	81160	95766	76569	11853	56920	36872	84068	23931	03132	98250	40881	36400
20655	61505	43434	77662	50397	04107	48089	33570	43315	01145	01259	41841	14027	91281	73123	73934	25643	73719	12961	91118
51220	32750	85161	17942	09500	82183	70192	61318	76271	83729	96120	66977	74568	21205	25466	97080	43938	06211	31215	83322
27999	50758	76499	08955	97396	68137	36721	50734	88856	55193	92580	17878	87290	67002	49445	09798	53583	18839	24688	27439
02835	40215	61818	64739	13109	61681	00418	26909	90229	36990	25826	20871	76561	91185	64162	90417	68072	39107	48467	74371
40953	38806	82384	00231	83815	30315	40698	38553	30566	62249	93172	84566	89662	26712	91300	53308	14136	07032	38650	22841
50731	92877	46395	86922	92330	33398	78200	33835	32614	81082	84756	01914	32359	27149	39812	24643	49913	43380	88439	19102
10959	25440	26269	40889	91641	78868	17601	76567	11357	01088	52233	21106	73798	90942	07779	42685	04186	61471	47687	20726
03784	41838	17267	04927	26719	30540	22557	33603	75689	04266	61592	18588	59135	95029	46711	01496	49891	22452	81489	62136
80949	08395	58909	64448	04736	07373	00130	08352	75058	58561	77656	67493	82480	03507	34742	82955	31274	17994	46276	01606
86038	76897	37132	44871	85577	07205	03919	19347	17449	86832	46996	84847	15684	15187	33558	28105	83358	15947	51285	01570
97916	32882	97441	26397	27173	46059	52260	76989	14728	68207	29811	11127	81957	79526	56240	35007	86620	23703	16099	46252
72451	18449	04444	30225	86543	30362	47162	45784	29045	26513	76680	75923	79273	43684	96519	86541	10836	10778	08017	82954
12623	20526	27902	28596	69351	73214	67953	43725	71702	07781	99830	83847	18818	94286	60973	57960	91843	86460	93269	35636
13305	23464	16745	59406	10177	27227	47841	74838	65382	63736	47603	65176	20206	25929	51398	80379	75345	50304	60320	31904
78104	00194	87152	34571	74435	35395	18567	65386	93855	40642	01960	26232	19832	04214	61808	92899	24707	22758	18685	56996
13593	59272	95778	69866	72803	98001	74976	28751	52090	22903	79050	55048	73203	95178	30158	52641	46841	20270	39583	50958
42926	75661	41312	82546	92060	17676	64499	68650	45971	98490	68982	38487	88558	03466	15752	38590	13687	63909	32355	47874
68071	11350	33669	51764	44213	16415	93085	95030	96409	98428	99776	77725	89102	00845	06364	31922	68229	02738	39651	68745
19602	77575	37169	65529	40604	17618	55960	21752	49454	15383	99010	99772	86920	32699	91168	32237	75433	93022	31898	72444

Each digit in the table was generated by a process equally likely to give any one of the ten digits 0, 1, 2, ..., 9. The digits have been grouped purely for convenience of reading.

Random digits can be used to simulate random samples from any probability distribution. First note that random numbers U from the continuous uniform distribution on (0:1) can be formed approximately by placing a decimal point in front of groups of, say, five random digits (again for ease of reading), thus: 0.02484, 0.88139, 0.31788, etc. These numbers may in turn be transformed to random numbers X from any continuous distribution with c.d.f. $F(x)$ by solving the equation $U = F(X)$ for X in terms of U – this may be accomplished using a graph of $F(x)$ as shown in the diagram. Random numbers from discrete distributions may be obtained by a similar graphical process or by finding the smallest X such that $F(X) > U$.

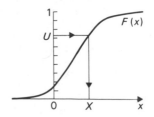

Random numbers from normal distributions

0.5117	−0.6501	−0.0240	−0.0374	0.4650	0.6573	−0.8489	−1.6237	0.9161	0.4286	2.1530	0.8024	0.6296	−0.7431	0.2311
0.4219	−0.1946	−0.2223	0.8529	0.3829	1.3436	1.4955	0.5792	−1.1305	−0.3346	−1.9110	1.4270	−1.7715	0.6190	1.3728
−0.3968	−2.0135	0.3052	1.4541	0.3063	0.0446	−2.1887	−0.2511	0.9978	−0.4531	−0.8269	−1.1302	−0.2418	0.1748	−0.2623
0.4687	−1.4781	−1.7345	0.7693	−0.9250	0.0144	0.7538	0.0476	−0.6648	1.0353	−1.9236	−0.0390	1.7233	−0.3012	−1.2579
0.6956	0.9457	−2.2365	−0.2212	−0.0329	1.3567	−1.0202	−0.6191	−1.5205	−2.4005	0.0528	−0.9080	−0.6263	0.6274	−0.1815
0.3644	1.5510	−0.4803	−1.0094	0.4757	0.9914	0.5532	−0.7414	0.6996	0.4086	−0.7131	0.5659	0.5726	−1.0370	0.6656
0.9069	−0.3967	0.6256	0.7654	0.6252	2.1284	1.2576	0.8842	0.3930	0.2474	−0.4700	0.5366	−0.7211	0.4170	−0.0039
−1.1476	−0.2261	−0.4645	0.3763	−1.5602	0.8831	1.4995	−0.5930	0.9010	0.5485	−0.8076	0.0739	1.8341	0.6792	−0.2652
0.6157	1.1829	−1.0711	−0.6905	0.2236	−0.4170	0.6114	0.0493	1.3242	1.0989	−1.3245	−0.0253	0.3983	1.7539	0.7943
−0.0140	0.3773	−1.0443	0.3281	−0.1657	−0.5163	0.0572	−1.7496	0.6925	−0.9631	2.6746	0.1739	−0.2046	−1.3770	−2.5394
0.6557	0.4607	−0.1899	1.4323	1.6818	−0.9194	−0.0812	−0.0136	0.5099	0.4716	0.4880	−1.2776	−0.5492	−0.7707	0.2670
1.2269	2.4441	−2.5492	−0.7248	−1.5706	−0.3898	−0.6462	1.5392	0.4541	−0.2495	−0.5361	−1.2611	0.1790	0.7144	−0.3908
−2.0647	−0.1562	−0.2500	1.2900	1.1793	0.4379	−0.5050	−0.8679	−0.2687	1.0452	−0.5523	1.2387	−1.6821	1.0840	−0.8673
0.2633	1.0436	0.3264	0.1131	−1.9656	0.2444	−0.4575	0.1475	−0.9912	−0.0698	1.4027	−1.4261	−1.3690	1.1719	0.6424
0.1536	−0.2625	−0.4261	0.1458	0.1283	−0.0728	1.0004	0.2144	1.7433	0.4577	−0.7605	−0.8476	−1.1592	3.0920	0.8802
0.0288	0.0438	−0.1742	0.9610	−0.3768	−0.1367	0.0709	0.7607	−1.2500	0.5741	1.6103	−0.1116	−0.3716	−1.3832	0.8992
−1.8426	−0.3121	−1.0415	0.5305	−0.9029	−0.9628	−0.3619	−0.9187	0.2634	−0.0089	−0.3599	0.8698	1.2590	−1.2478	−0.8828
−0.7422	−0.5728	0.6748	1.9620	−0.0364	0.3374	0.6351	1.7987	−0.0415	0.9141	−0.7215	−0.6227	1.1671	−1.0297	0.5019
−0.8158	1.6473	−2.0569	−0.5147	0.5564	−1.0821	−1.7388	0.0251	−1.3612	−2.2882	0.3054	−1.2463	1.3680	0.1380	1.5723
1.2816	0.4435	0.3760	−0.6307	0.9982	1.9737	−0.1486	0.5829	1.7779	0.8335	−0.4614	0.7387	−0.9224	1.4158	0.4807
0.3257	1.6609	1.5465	1.8711	0.4291	−0.4098	−0.9554	0.5928	0.6828	2.8234	0.7119	0.2455	−0.2270	−0.9025	0.1486
−0.5662	0.2938	−1.0305	0.4343	2.1240	1.5033	−0.5762	1.0887	−0.0615	−1.4243	0.9548	−1.2092	−0.1559	0.8749	−0.1916
−0.7432	0.6906	−1.9848	−0.2062	1.5273	1.1176	−0.4626	−1.7566	−0.2784	0.3495	−0.4353	−2.5354	−1.8229	−1.2539	−0.5565
0.0799	0.8198	−1.2491	0.4998	−0.0589	−0.6848	−0.9974	0.8797	−0.0676	1.0889	−0.5973	−3.1585	0.4271	0.6168	2.1738
0.7719	1.2595	−0.1923	−1.8775	1.2376	−0.4795	−0.6284	−0.0667	−0.5308	−0.2933	0.7285	−1.6920	−1.7669	0.5144	−0.5109

These random numbers are from the standard normal distribution, i.e. the normal distribution with mean 0 and standard deviation 1. They may be transformed to random numbers from any other normal distribution with mean μ and standard deviation σ by multiplying them by σ and adding μ. For example, to obtain a sample from the normal distribution with mean $\mu = 10$ and standard deviation $\sigma = 2$, double the numbers and add 10, thus: $2 \times (0.5117) + 10 = 11.0234$, $2 \times (−0.6501) + 10 = 8.6998$, $2 \times (−0.0240) + 10 = 9.9520$, etc.

Random numbers from exponential distributions

0.6193	1.8350	0.2285	1.5106	0.5024	2.3326	4.7123	0.9869	0.7543	0.1759	2.3678	0.1260	1.5913	0.1730	0.5110
0.0354	1.4300	1.6249	0.1402	0.8824	0.9866	0.2289	0.1741	1.3838	0.3772	1.5610	0.1928	0.6389	0.1052	0.4661
0.1258	0.2010	0.2728	0.5152	1.2431	0.3924	1.4429	0.5880	0.0941	1.9999	0.2395	2.6969	1.5680	3.7064	0.0875
2.0308	1.0043	0.1779	0.2475	0.2649	0.2800	5.0992	2.2468	2.2083	0.0988	0.0611	2.2454	0.9630	0.8355	4.0204
0.2145	2.5019	1.3019	1.6369	1.3499	0.6203	1.9118	0.1670	0.1949	1.3440	0.2005	1.5157	1.7353	0.9324	1.3523
1.1118	1.9728	0.6191	0.0149	0.5376	0.0046	0.6752	1.6281	0.2772	0.0556	0.4470	0.5266	0.8817	0.2427	1.1638
0.2432	0.7302	2.4396	0.0779	1.0151	0.4888	1.2114	0.3606	0.0234	1.9367	1.2689	2.1829	0.3569	1.4470	0.9422
0.6834	1.2602	0.0440	3.6550	0.1032	1.5326	4.1297	1.2753	0.1516	0.3470	0.9681	0.4149	1.5600	1.7575	0.5968
0.8743	0.5972	0.5226	0.6086	0.4820	0.8126	0.7244	2.8622	1.2995	0.1391	1.0467	0.3153	0.7654	0.0526	0.6286
1.8945	0.0828	0.6279	0.5823	1.7757	0.1087	0.6876	0.5346	0.6817	0.1436	0.6388	0.6211	0.8468	0.9272	0.8470
1.6711	0.2592	2.1458	0.0449	3.1336	0.5581	0.1607	0.4598	0.7907	0.5938	2.7818	1.8210	1.2763	1.2032	0.0126
0.5536	0.3020	0.2853	1.2290	0.4552	0.0068	1.5726	0.0027	0.0645	0.2775	3.1438	2.9250	0.8723	4.8510	1.2586
0.9866	0.9132	0.3053	0.3737	0.5469	0.0346	2.8317	0.2933	0.7938	0.2877	0.2119	0.8928	2.0636	0.5153	0.8829
1.3695	0.2366	1.7697	1.0209	0.7348	2.3026	0.0673	1.2728	0.5977	5.5840	1.0013	0.4362	0.4095	1.7154	0.0811
0.5208	0.6984	1.0987	0.1917	0.6229	2.1011	0.0072	1.4618	1.1227	0.6920	0.3934	1.3236	0.2127	0.1735	1.0092
2.2593	4.3931	1.4765	0.7746	2.6811	0.0104	0.4500	0.2286	0.1451	0.2324	0.6069	1.2613	1.9487	1.2471	1.3712
1.0490	0.5225	0.2698	0.6562	0.3095	0.7785	0.3197	0.6824	0.3432	0.4526	2.7164	1.0550	0.6933	1.8137	1.7805
0.0518	0.3456	0.1365	0.4320	4.4838	1.1652	0.0927	0.7937	0.0223	1.4675	0.1545	1.4515	0.8765	0.1045	0.2226
0.7941	0.3201	0.0899	1.6611	0.5771	0.2266	0.3686	0.0393	0.8588	0.4303	0.4266	0.3845	0.5723	2.6542	0.6612
0.4676	0.5834	2.3247	0.7372	2.4606	0.3932	0.1851	1.6538	1.7101	1.4550	0.4140	0.0591	0.8581	3.3141	0.4378
0.9766	0.8192	4.1140	0.5508	0.3703	2.3148	0.0545	1.3626	0.3847	2.1840	3.6072	0.1066	0.7252	1.3741	0.8290
1.2443	0.5925	2.2355	0.1753	0.4353	0.7177	3.4943	0.8487	3.9863	2.8398	2.2733	0.4179	0.5265	1.6294	0.4912
0.6793	0.3157	1.6361	0.7469	2.5568	0.2092	0.0555	2.0506	0.1296	1.9426	0.0250	0.9036	1.3022	0.4394	0.6579
0.2690	0.4206	0.9004	2.7633	0.2804	2.7984	2.5987	0.1178	0.5429	1.6306	3.0790	1.1955	0.0738	0.1938	2.0874
0.2610	0.1912	0.3160	1.1692	2.8068	0.2948	0.1969	1.3823	2.1179	0.3821	1.8986	1.3541	0.1657	4.3879	3.3662

These are random numbers from the exponential distribution with mean 1. They may be transformed to random numbers from any other exponential distribution with mean μ simply by multiplying them by μ. Thus a sample from the exponential distribution with mean 10 is 6.193, 18.350, 2.285, . . . , etc.

Binomial coefficients

$$\binom{n}{r} = \binom{n}{n-r} = \frac{n!}{r!(n-r)!} = \frac{n(n-1)\ldots(n-r+1)}{r(r-1)\ldots 1}$$

for $n = 1$ to 36 and 52 (for playing-card problems)

n \ r	0	1	2	3	4	5	6	7	8	9	10	11
1	1	1										
2	1	2	1									
3	1	3	3	1								
4	1	4	6	4	1							
5	1	5	10	10	5	1						
6	1	6	15	20	15	6	1					
7	1	7	21	35	35	21	7	1				
8	1	8	28	56	70	56	28	8	1			
9	1	9	36	84	126	126	84	36	9	1		
10	1	10	45	120	210	252	210	120	45	10	1	
11	1	11	55	165	330	462	462	330	165	55	11	1
12	1	12	66	220	495	792	924	792	495	220	66	12
13	1	13	78	286	715	1287	1716	1716	1287	715	286	78
14	1	14	91	364	1001	2002	3003	3432	3003	2002	1001	364
15	1	15	105	455	1365	3003	5005	6435	6435	5005	3003	1365
16	1	16	120	560	1820	4368	8008	11440	12870	11440	8008	4368
17	1	17	136	680	2380	6188	12376	19448	24310	24310	19448	12376
18	1	18	153	816	3060	8568	18564	31824	43758	48620	43758	31824
19	1	19	171	969	3876	11628	27132	50388	75582	92378	92378	75582
20	1	20	190	1140	4845	15504	38760	77520	125970	167960	184756	167960
21	1	21	210	1330	5985	20349	54264	116280	203490	293930	352716	352716
22	1	22	231	1540	7315	26334	74613	170544	319770	497420	646646	705432
23	1	23	253	1771	8855	33649	100947	245157	490314	817190	1144066	1352078
24	1	24	276	2024	10626	42504	134596	346104	735471	1307504	1961256	2496144
25	1	25	300	2300	12650	53130	177100	480700	1081575	2042975	3268760	4457400
26	1	26	325	2600	14950	65780	230230	657800	1562275	3124550	5311735	7726160
27	1	27	351	2925	17550	80730	296010	888030	2220075	4686825	8436285	13037895
28	1	28	378	3276	20475	98280	376740	1184040	3108105	6906900	13123110	21474180
29	1	29	406	3654	23751	118755	475020	1560780	4292145	10015005	20030010	34597290
30	1	30	435	4060	27405	142506	593775	2035800	5852925	14307150	30045015	54627300
31	1	31	465	4495	31465	169911	736281	2629575	7888725	20160075	44352165	84672315
32	1	32	496	4960	35960	201376	906192	3365856	10518300	28048800	64512240	129024480
33	1	33	528	5456	40920	237336	1107568	4272048	13884156	38567100	92561040	193536720
34	1	34	561	5984	46376	278256	1344904	5379616	18156204	52451256	131128140	286097760
35	1	35	595	6545	52360	324632	1623160	6724520	23535820	70607460	183579396	417225900
36	1	36	630	7140	58905	376992	1947792	8347680	30260340	94143280	254186856	600805296
52	1	52	1326	22100	270725	2598960	20358520	133784560	752538150	3679075400	15820024220	60403728840

n \ r	12	13	14	15	16	17	18
12	1						
13	13	1					
14	91	14	1				
15	455	105	15	1			
16	1820	560	120	16	1		
17	6188	2380	680	136	17	1	
18	18564	8568	3060	816	153	18	1
19	50388	27132	11628	3876	969	171	19
20	125970	77520	38760	15504	4845	1140	190
21	293930	203490	116280	54264	20349	5985	1330
22	646646	497420	319770	170544	74613	26334	7315
23	1352078	1144066	817190	490314	245157	100947	33649
24	2704156	2496144	1961256	1307504	735471	346104	134596
25	5200300	5200300	4457400	3268760	2042975	1081575	480700
26	9657700	10400600	9657700	7726160	5311735	3124550	1562275
27	17383860	20058300	20058300	17383860	13037895	8436285	4686825
28	30421755	37442160	40116600	37442160	30421755	21474180	13123110
29	51895935	67863915	77558760	77558760	67863915	51895935	34597290
30	86493225	119759850	145422675	155117520	145422675	119759850	86493225
31	141120525	206253075	265182525	300540195	300540195	265182525	206253075
32	225792840	347373600	471435600	565722720	601080390	565722720	471435600
33	354817320	573166440	818809200	1037158320	1166803110	1166803110	1037158320
34	548354040	927983760	1391975640	1855967520	2203961430	2333606220	2203961430
35	834451800	1476337800	2319959400	3247943160	4059928950	4537567650	4537567650
36	1251677700	2310789600	3796297200	5567902560	7307872110	8597496600	9075135300
52	206379406870	635013559600	1768966344600	4481381406320	10363194502115	21945588357420	42671977361650

n = 52	r		r		r		r	
	19	76360380541900	20	125994627894135	21	191991813933920	22	270533919634160
	23	352870329957600	24	426384982032100	25	477551179875952	26	495918532948104

The binomial coefficient $\binom{n}{r}$ gives the number of different groups of r objects which may be selected from a collection of n objects: e.g. there are $\binom{4}{2} = 6$ different pairs of letters which may be selected from the four letters A, B, C, D; they are (A,B), (A,C), (A,D), (B,C), (B,D) and (C,D). The *order* of selection is presumed immaterial, so (B,A) is regarded as the same as (A,B) etc. As a more substantial example, the number of different hands of five cards which may be dealt from a full pack of 52 cards is $\binom{52}{5} = 2\,598\,960$.

Reciprocals, squares, square roots and their reciprocals, and factorials

n	1/n	n^2	$\sqrt{n}$	$\sqrt{10n}$	$1/\sqrt{n}$	$1/\sqrt{10n}$	n!
1	1.0000	1	1.0000	3.1623	1.000	.31623	1
2	.50000	4	1.4142	4.4721	.7071	.22361	2
3	.33333	9	1.7321	5.4772	.5774	.18257	6
4	.25000	16	2.0000	6.3246	.5000	.15811	24
5	.20000	25	2.2361	7.0711	.4472	.14142	120
6	.16667	36	2.4495	7.7460	.4082	.12910	720
7	.14286	49	2.6458	8.3666	.3780	.11952	5,040
8	.12500	64	2.8284	8.9443	.3536	.11180	40,320
9	.11111	81	3.0000	9.4868	.3333	.10541	362,880
10	.10000	100	3.1623	10.000	.3162	.10000	3,628,800
11	.09091	121	3.3166	10.488	.3015	.09535	39,916,800
12	.08333	144	3.4641	10.954	.2887	.09129	4.7900×10^{8}
13	.07692	169	3.6056	11.402	.2774	.08771	6.2270×10^{9}
14	.07143	196	3.7417	11.832	.2673	.08452	8.7178×10^{10}
15	.06667	225	3.8730	12.247	.2582	.08165	1.3077×10^{12}
16	.06250	256	4.0000	12.649	.2500	.07906	2.0923×10^{13}
17	.05882	289	4.1231	13.038	.2425	.07670	3.5569×10^{14}
18	.05556	324	4.2426	13.416	.2357	.07454	6.4024×10^{15}
19	.05263	361	4.3589	13.784	.2294	.07255	1.2165×10^{17}
20	.05000	400	4.4721	14.142	.2236	.07071	2.4329×10^{18}
21	.04762	441	4.5826	14.491	.2182	.06901	5.1091×10^{19}
22	.04545	484	4.6904	14.832	.2132	.06742	1.1240×10^{21}
23	.04348	529	4.7958	15.166	.2085	.06594	2.5852×10^{22}
24	.04167	576	4.8990	15.492	.2041	.06455	6.2045×10^{23}
25	.04000	625	5.0000	15.811	.2000	.06325	1.5511×10^{25}
26	.03846	676	5.0990	16.125	.1961	.06202	4.0329×10^{26}
27	.03704	729	5.1962	16.432	.1925	.06086	1.0889×10^{28}
28	.03571	784	5.2915	16.733	.1890	.05976	3.0489×10^{29}
29	.03448	841	5.3852	17.029	.1857	.05872	8.8418×10^{30}
30	.03333	900	5.4772	17.321	.1826	.05774	2.6525×10^{32}
31	.03226	961	5.5678	17.607	.1796	.05680	8.2228×10^{33}
32	.03125	1024	5.6569	17.889	.1768	.05590	2.6313×10^{35}
33	.03030	1089	5.7446	18.166	.1741	.05505	8.6833×10^{36}
34	.02941	1156	5.8310	18.439	.1715	.05423	2.9523×10^{38}
35	.02857	1225	5.9161	18.708	.1690	.05345	1.0333×10^{40}
36	.02778	1296	6.0000	18.974	.1667	.05270	3.7199×10^{41}
37	.02703	1369	6.0828	19.235	.1644	.05199	1.3764×10^{43}
38	.02632	1444	6.1644	19.494	.1622	.05130	5.2302×10^{44}
39	.02564	1521	6.2450	19.748	.1601	.05064	2.0398×10^{46}
40	.02500	1600	6.3246	20.000	.1581	.05000	8.1592×10^{47}
41	.02439	1681	6.4031	20.248	.1562	.04939	3.3453×10^{49}
42	.02381	1764	6.4807	20.494	.1543	.04880	1.4050×10^{51}
43	.02326	1849	6.5574	20.736	.1525	.04822	6.0415×10^{52}
44	.02273	1936	6.6332	20.976	.1508	.04767	2.6583×10^{54}
45	.02222	2025	6.7082	21.213	.1491	.04714	1.1962×10^{56}
46	.02174	2116	6.7823	21.448	.1474	.04663	5.5026×10^{57}
47	.02128	2209	6.8557	21.679	.1459	.04613	2.5862×10^{59}
48	.02083	2304	6.9282	21.909	.1443	.04564	1.2414×10^{61}
49	.02041	2401	7.0000	22.136	.1429	.04518	6.0828×10^{62}
50	.02000	2500	7.0711	22.361	.1414	.04472	3.0414×10^{64}

n	1/n	n^2	$\sqrt{n}$	$\sqrt{10n}$	$1/\sqrt{n}$	$1/\sqrt{10n}$	n!
51	.01961	2601	7.1414	22.583	.1400	.04428	1.5511×10^{66}
52	.01923	2704	7.2111	22.804	.1387	.04385	8.0658×10^{67}
53	.01887	2809	7.2801	23.022	.1374	.04344	4.2749×10^{69}
54	.01852	2916	7.3485	23.238	.1361	.04303	2.3084×10^{71}
55	.01818	3025	7.4162	23.452	.1348	.04264	1.2696×10^{73}
56	.01786	3136	7.4833	23.664	.1336	.04226	7.1100×10^{74}
57	.01754	3249	7.5498	23.875	.1325	.04189	4.0527×10^{76}
58	.01724	3364	7.6158	24.083	.1313	.04152	2.3506×10^{78}
59	.01695	3481	7.6811	24.290	.1302	.04117	1.3868×10^{80}
60	.01667	3600	7.7460	24.495	.1291	.04082	8.3210×10^{81}
61	.01639	3721	7.8102	24.698	.1280	.04049	5.0758×10^{83}
62	.01613	3844	7.8740	24.900	.1270	.04016	3.1470×10^{85}
63	.01587	3969	7.9373	25.100	.1260	.03984	1.9826×10^{87}
64	.01563	4096	8.0000	25.298	.1250	.03953	1.2689×10^{89}
65	.01538	4225	8.0623	25.495	.1240	.03922	8.2477×10^{90}
66	.01515	4356	8.1240	25.690	.1231	.03892	5.4434×10^{92}
67	.01493	4489	8.1854	25.884	.1222	.03863	3.6471×10^{94}
68	.01471	4624	8.2462	26.077	.1213	.03835	2.4800×10^{96}
69	.01449	4761	8.3066	26.268	.1204	.03807	1.7112×10^{98}
70	.01429	4900	8.3666	26.458	.1195	.03780	1.1979×10^{100}
71	.01408	5041	8.4261	26.646	.1187	.03753	8.5048×10^{101}
72	.01389	5184	8.4853	26.833	.1179	.03727	6.1234×10^{103}
73	.01370	5329	8.5440	27.019	.1170	.03701	4.4701×10^{105}
74	.01351	5476	8.6023	27.203	.1162	.03676	3.3079×10^{107}
75	.01333	5625	8.6603	27.386	.1155	.03651	2.4809×10^{109}
76	.01316	5776	8.7178	27.568	.1147	.03627	1.8855×10^{111}
77	.01299	5929	8.7750	27.749	.1140	.03604	1.4518×10^{113}
78	.01282	6084	8.8318	27.928	.1132	.03581	1.1324×10^{115}
79	.01266	6241	8.8882	28.107	.1125	.03558	8.9462×10^{116}
80	.01250	6400	8.9443	28.284	.1118	.03536	7.1569×10^{118}
81	.01235	6561	9.0000	28.460	.1111	.03514	5.7971×10^{120}
82	.01220	6724	9.0554	28.636	.1104	.03492	4.7536×10^{122}
83	.01205	6889	9.1104	28.810	.1098	.03471	3.9455×10^{124}
84	.01190	7056	9.1652	28.983	.1091	.03450	3.3142×10^{126}
85	.01176	7225	9.2195	29.155	.1085	.03430	2.8171×10^{128}
86	.01163	7396	9.2736	29.326	.1078	.03410	2.4227×10^{130}
87	.01149	7569	9.3274	29.496	.1072	.03390	2.1078×10^{132}
88	.01136	7744	9.3808	29.665	.1066	.03371	1.8548×10^{134}
89	.01124	7921	9.4340	29.833	.1060	.03352	1.6508×10^{136}
90	.01111	8100	9.4868	30.000	.1054	.03333	1.4857×10^{138}
91	.01099	8281	9.5394	30.166	.1048	.03315	1.3520×10^{140}
92	.01087	8464	9.5917	30.332	.1043	.03297	1.2438×10^{142}
93	.01075	8649	9.6437	30.496	.1037	.03279	1.1568×10^{144}
94	.01064	8836	9.6954	30.659	.1031	.03262	1.0874×10^{146}
95	.01053	9025	9.7468	30.822	.1026	.03244	1.0330×10^{148}
96	.01042	9216	9.7980	30.984	.1021	.03227	9.9168×10^{149}
97	.01031	9409	9.8489	31.145	.1015	.03211	9.6193×10^{151}
98	.01020	9604	9.8995	31.305	.1010	.03194	9.4269×10^{153}
99	.01010	9801	9.9499	31.464	.1005	.03178	9.3326×10^{155}
100	.01000	10000	10.000	31.623	.1000	.03162	9.3326×10^{157}

Useful constants

π	3.14159 26536	$1/\pi$	0.31830 98862	π^2	9.86960 44011	$1/\pi^2$	0.10132 11836
$\sqrt{\pi}$	1.77245 38509	$1/\sqrt{\pi}$	0.56418 95835	$\sqrt{2\pi}$	2.50662 82746	$1/\sqrt{2\pi}$	0.39894 22804
e	2.71828 18285	$1/e$	0.36787 94412	$\sqrt{e}$	1.64872 12707	$1/\sqrt{e}$	0.60653 06597
$\log_e 10$	2.30258 50930	$\log_{10} e$	0.43429 44819	$\log_e \pi$	1.14472 98858	$\log_{10} \pi$	0.49714 98727
$\sqrt{2}$	1.41421 35624	$1/\sqrt{2}$	0.70710 67812	$\sqrt{3}$	1.73205 08076	$1/\sqrt{3}$	0.57735 02692

The negative exponential function: e^-x

x	0	1	2	3	4	5	6	7	8	9
0.0	1.0000	9900	9802	9704	9608	9512	9418	9324	9231	9139
0.1	9048	8958	8869	8781	8694	8607	8521	8437	8353	8270
0.2	8187	8106	8025	7945	7866	7788	7711	7634	7558	7483
0.3	7408	7334	7261	7189	7118	7047	6977	6907	6839	6771
0.4	6703	6637	6570	6505	6440	6376	6313	6250	6188	6126
0.5	6065	6005	5945	5886	5827	5769	5712	5655	5599	5543
0.6	5488	5434	5379	5326	5273	5220	5169	5117	5066	5016
0.7	4966	4916	4868	4819	4771	4724	4677	4630	4584	4538
0.8	4493	4449	4404	4360	4317	4274	4232	4190	4148	4107
0.9	4066	4025	3985	3946	3906	3867	3829	3791	3753	3716
1.0	3679	3642	3606	3570	3535	3499	3465	3430	3396	3362
1.1	3329	3296	3263	3230	3198	3166	3135	3104	3073	3042
1.2	3012	2982	2952	2923	2894	2865	2837	2808	2780	2753
1.3	2725	2698	2671	2645	2618	2592	2567	2541	2516	2491
1.4	2466	2441	2417	2393	2369	2346	2322	2299	2276	2254
1.5	2231	2209	2187	2165	2144	2122	2101	2080	2060	2039
1.6	2019	1999	1979	1959	1940	1920	1901	1882	1864	1845
1.7	1827	1809	1791	1773	1755	1738	1720	1703	1686	1670
1.8	1653	1637	1620	1604	1588	1572	1557	1541	1526	1511
1.9	1496	1481	1466	1451	1437	1423	1409	1395	1381	1367
2.0	1353	1340	1327	1313	1300	1287	1275	1262	1249	1237
2.1	1225	1212	1200	1188	1177	1165	1153	1142	1130	1119
2.2	1108	1097	1086	1075	1065	1054	1044	1033	1023	1013
2.3	1003	0993	0983	0973	0963	0954	0944	0935	0926	0916
2.4	0907	0898	0889	0880	0872	0863	0854	0846	0837	0829
2.5	0821	0813	0805	0797	0789	0781	0773	0765	0758	0750
2.6	0743	0735	0728	0721	0714	0707	0699	0693	0686	0679
2.7	0672	0665	0659	0652	0646	0639	0633	0627	0620	0614
2.8	0608	0602	0596	0590	0584	0578	0573	0567	0561	0556
2.9	0550	0545	0539	0534	0529	0523	0518	0513	0508	0503
3.0	0498	0493	0488	0483	0478	0474	0469	0464	0460	0455
3.1	0450	0446	0442	0437	0433	0429	0424	0420	0416	0412
3.2	0408	0404	0400	0396	0392	0388	0384	0380	0376	0373
3.3	0369	0365	0362	0358	0354	0351	0347	0344	0340	0337
3.4	0334	0330	0327	0324	0321	0317	0314	0311	0308	0305
3.5	0302	0299	0296	0293	0290	0287	0284	0282	0279	0276
3.6	0273	0271	0268	0265	0263	0260	0257	0255	0252	0250
3.7	0247	0245	0242	0240	0238	0235	0233	0231	0228	0226
3.8	0224	0221	0219	0217	0215	0213	0211	0209	0207	0204
3.9	0202	0200	0198	0196	0194	0193	0191	0189	0187	0185

SUBTRACT PROPORTIONAL PARTS

x	1	2	3	4	5	6	7	8	9
0.0	10	20	29	39	49	59	68	78	88
0.1	9	19	28	37	47	56	65	74	84
0.2	9	18	27	35	44	53	62	71	79
0.3	8	17	25	34	42	50	59	67	76
0.4	8	16	24	32	40	48	56	64	72
0.5	6	12	17	23	29	35	40	46	52
0.6	5	10	16	21	26	31	37	42	47
0.7	5	9	14	19	24	28	33	38	43
0.8	4	9	13	17	22	26	30	34	39
0.9	4	8	12	16	19	23	27	31	35
1.0	4	7	11	14	18	21	25	28	32
1.1	3	6	10	13	16	19	22	25	29
1.2	3	6	9	11	14	17	20	23	26
1.3	3	5	8	10	13	16	18	21	23
1.4	2	5	7	9	12	14	16	19	21
1.5	2	4	6	9	11	13	15	17	19
1.6	2	4	6	8	10	12	13	15	17
1.7	2	3	5	7	8	10	12	14	16
1.8	2	3	5	6	8	9	11	13	14
1.9	1	3	4	6	7	9	10	11	13

The exponential function: e^x

x	0	1	2	3	4	5	6	7	8	9
0.0	1.0000	1.0101	1.0202	1.0305	1.0408	1.0513	1.0618	1.0725	1.0833	1.0942
0.1	1.1052	1.1163	1.1275	1.1388	1.1503	1.1618	1.1735	1.1853	1.1972	1.2092
0.2	1.2214	1.2337	1.2461	1.2586	1.2712	1.2840	1.2969	1.3100	1.3231	1.3364
0.3	1.3499	1.3634	1.3771	1.3910	1.4049	1.4191	1.4333	1.4477	1.4623	1.4770
0.4	1.4918	1.5068	1.5220	1.5373	1.5527	1.5683	1.5841	1.6000	1.6161	1.6323
0.5	1.6487	1.6653	1.6820	1.6989	1.7160	1.7333	1.7507	1.7683	1.7860	1.8040
0.6	1.8221	1.8404	1.8589	1.8776	1.8965	1.9155	1.9348	1.9542	1.9739	1.9937
0.7	2.0138	2.0340	2.0544	2.0751	2.0959	2.1170	2.1383	2.1598	2.1815	2.2034
0.8	2.2255	2.2479	2.2705	2.2933	2.3164	2.3396	2.3632	2.3869	2.4109	2.4351
0.9	2.4596	2.4843	2.5093	2.5345	2.5600	2.5857	2.6117	2.6379	2.6645	2.6912
1.0	2.7183	2.7456	2.7732	2.8011	2.8292	2.8577	2.8864	2.9154	2.9447	2.9743
1.1	3.0042	3.0344	3.0649	3.0957	3.1268	3.1582	3.1899	3.2220	3.2544	3.2871
1.2	3.3201	3.3535	3.3872	3.4212	3.4556	3.4903	3.5254	3.5609	3.5966	3.6328
1.3	3.6693	3.7062	3.7434	3.7810	3.8190	3.8574	3.8962	3.9354	3.9749	4.0149
1.4	4.0552	4.0960	4.1371	4.1787	4.2207	4.2631	4.3060	4.3492	4.3929	4.4371
1.5	4.4817	4.5267	4.5722	4.6182	4.6646	4.7115	4.7588	4.8066	4.8550	4.9037
1.6	4.9530	5.0028	5.0531	5.1039	5.1552	5.2070	5.2593	5.3122	5.3656	5.4195
1.7	5.4739	5.5290	5.5845	5.6407	5.6973	5.7546	5.8124	5.8709	5.9299	5.9896
1.8	6.0496	6.1104	6.1719	6.2339	6.2965	6.3598	6.4237	6.4883	6.5535	6.6194
1.9	6.6859	6.7531	6.8210	6.8895	6.9588	7.0287	7.0993	7.1707	7.2427	7.3155
2.0	7.3891	7.4633	7.5383	7.6141	7.6906	7.7679	7.8460	7.9248	8.0045	8.0849
2.1	8.1662	8.2482	8.3311	8.4149	8.4994	8.5849	8.6711	8.7583	8.8463	8.9352
2.2	9.0250	9.1157	9.2073	9.2999	9.3933	9.4877	9.5831	9.6794	9.7767	9.8749
2.3	9.9742	10.074	10.176	10.278	10.381	10.486	10.591	10.697	10.805	10.913
2.4	11.023	11.134	11.246	11.359	11.473	11.588	11.705	11.822	11.941	12.061
2.5	12.182	12.305	12.429	12.554	12.680	12.807	12.936	13.066	13.197	13.330
2.6	13.464	13.599	13.736	13.874	14.013	14.154	14.296	14.440	14.585	14.732
2.7	14.880	15.029	15.180	15.333	15.487	15.643	15.800	15.959	16.119	16.281
2.8	16.445	16.610	16.777	16.945	17.116	17.288	17.462	17.637	17.814	17.993
2.9	18.174	18.357	18.541	18.728	18.916	19.106	19.298	19.492	19.688	19.886
3.0	20.086	20.287	20.491	20.697	20.905	21.115	21.328	21.542	21.758	21.977
3.1	22.198	22.421	22.646	22.874	23.104	23.336	23.571	23.807	24.047	24.288
3.2	24.533	24.779	25.028	25.280	25.534	25.790	26.050	26.311	26.576	26.843
3.3	27.113	27.385	27.660	27.938	28.219	28.503	28.789	29.079	29.371	29.666
3.4	29.964	30.265	30.569	30.877	31.187	31.500	31.817	32.137	32.460	32.786
3.5	33.115	33.448	33.784	34.124	34.467	34.813	35.163	35.517	35.874	36.234
3.6	36.598	36.966	37.338	37.713	38.092	38.475	38.861	39.252	39.646	40.045
3.7	40.447	40.854	41.264	41.679	42.098	42.521	42.948	43.380	43.816	44.256
3.8	44.701	45.150	45.604	46.063	46.525	46.993	47.465	47.942	48.424	48.911
3.9	49.402	49.899	50.400	50.907	51.419	51.935	52.457	52.985	53.517	54.055
4.0	54.598	55.147	55.701	56.261	56.826	57.397	57.974	58.557	59.145	59.740
4.1	60.340	60.947	61.559	62.178	62.803	63.434	64.072	64.715	65.366	66.023
4.2	66.686	67.357	68.033	68.717	69.408	70.105	70.810	71.522	72.240	72.966
4.3	73.700	74.440	75.189	75.944	76.708	77.478	78.257	79.044	79.838	80.640
4.4	81.451	82.269	83.096	83.931	84.775	85.627	86.488	87.357	88.235	89.121
4.5	90.017	90.922	91.836	92.759	93.691	94.632	95.583	96.544	97.514	98.494
4.6	99.484	100.48	101.49	102.51	103.54	104.58	105.64	106.70	107.77	108.85
4.7	109.95	111.05	112.17	113.30	114.43	115.58	116.75	117.92	119.10	120.30
4.8	121.51	122.73	123.97	125.21	126.47	127.74	129.02	130.32	131.63	132.95
4.9	134.29	135.64	137.00	138.38	139.77	141.17	142.59	144.03	145.47	146.94
5.	148.41	164.02	181.27	200.34	221.41	244.69	270.43	298.87	330.30	365.04
6.	403.43	445.86	492.75	544.57	601.85	665.14	735.10	812.41	897.85	992.27
7.	1096.6	1212.0	1339.4	1480.3	1636.0	1808.0	1998.2	2208.3	2440.6	2697.3
8.	2981.0	3294.5	3641.0	4023.9	4447.1	4914.8	5431.7	6002.9	6634.2	7332.0
9.	8103.1	8955.3	9897.1	10938	12088	13360	14765	16318	18034	19930
10.	22026	24343	26903	29733	32860	36316	40135	44356	49021	54176

The negative exponential function: e^-x (large x)

x	0	1	2	3	4	5	6	7	8	9
4.	0,018316	.016573	.014996	.013569	.012277	.011109	.010052	.009095	.008230	.007447
5.	.006738	.006097	.005517	.004992	.004517	.004087	.003698	.003346	.003028	.002739
6.	.002479	.002243	.002029	.001836	.001662	.001503	.001360	.001231	.001114	.001008
7.	$0^3$9119	$0^3$8251	$0^3$7466	$0^3$6755	$0^3$6113	$0^3$5531	$0^3$5005	$0^3$4528	$0^3$4097	$0^3$3707
8.	$0^3$3355	$0^3$3035	$0^3$2747	$0^3$2485	$0^3$2249	$0^3$2035	$0^3$1841	$0^3$1666	$0^3$1507	$0^3$1364
9.	$0^4$1234	$0^4$1117	$0^4$1010	$0^3$9142	$0^3$8272	$0^3$7485	$0^3$6773	$0^3$6128	$0^3$5545	$0^3$5017
10.	$0^4$4540	$0^4$4108	$0^4$3717	$0^4$3363	$0^4$3043	$0^4$2754	$0^4$2492	$0^4$2254	$0^4$2040	$0^4$1846

SUBTRACT PROPORTIONAL PARTS

Natural logarithms: $\log_e x$ or $\ln x$

x	0	1	2	3	4	5	6	7	8	9	1	2	3	4	5	6	7	8	9
											ADD PROPORTIONAL PARTS								
1.0	0.0000	0100	0198	0296	0392	0488	0583	0677	0770	0862	10	20	29	39	49	59	68	78	88
1.1	0.0953	1044	1133	1222	1310	1398	1484	1570	1655	1740	9	18	27	36	45	53	62	71	80
1.2	0.1823	1906	1989	2070	2151	2231	2311	2390	2469	2546	8	16	25	33	41	49	57	65	74
1.3	0.2624	2700	2776	2852	2927	3001	3075	3148	3221	3293	8	15	23	30	38	45	53	60	68
1.4	0.3365	3436	3507	3577	3646	3716	3784	3853	3920	3988	7	14	22	28	35	42	49	56	65
1.5	0.4055	4121	4187	4253	4318	4383	4447	4511	4574	4637	7	13	20	26	33	39	46	52	59
1.6	0.4700	4762	4824	4886	4947	5008	5068	5128	5188	5247	6	12	18	25	31	37	43	49	55
1.7	0.5306	5365	5423	5481	5539	5596	5653	5710	5766	5822	6	12	17	23	29	35	40	46	52
1.8	0.5878	5933	5988	6043	6098	6152	6206	6259	6313	6366	5	11	16	22	27	33	38	44	49
1.9	0.6419	6471	6523	6575	6627	6678	6729	6780	6831	6881	5	10	16	21	26	31	37	42	47
2.0	0.6931	6981	7031	7080	7129	7178	7227	7275	7324	7372	5	10	15	20	24	29	34	39	44
2.1	0.7419	7467	7514	7561	7608	7655	7701	7747	7793	7839	5	9	14	19	23	28	33	37	42
2.2	0.7885	7930	7975	8020	8065	8109	8154	8198	8242	8286	4	9	13	18	22	27	31	36	40
2.3	0.8329	8372	8416	8459	8502	8544	8587	8629	8671	8713	4	9	13	17	21	26	30	34	38
2.4	0.8755	8796	8838	8879	8920	8961	9002	9042	9083	9123	4	8	12	16	20	25	29	33	37
2.5	0.9163	9203	9243	9282	9322	9361	9400	9439	9478	9517	4	8	12	16	20	24	27	31	35
2.6	0.9555	9594	9632	9670	9708	9746	9783	9821	9858	9895	4	8	11	15	19	23	26	30	34
2.7	0.9933	9969	0006	0043	0080	0116	0152	0188	0225	0260	4	7	11	15	18	22	25	29	33
2.8	1.0296	0332	0367	0403	0438	0473	0508	0543	0578	0613	4	7	11	14	18	21	25	28	32
2.9	1.0647	0682	0716	0750	0784	0818	0852	0886	0919	0953	3	7	10	14	17	20	24	27	31
3.0	1.0986	1019	1053	1086	1119	1151	1184	1217	1249	1282	3	7	10	13	16	20	23	26	30
3.1	1.1314	1346	1378	1410	1442	1474	1506	1537	1569	1600	3	6	10	13	16	19	22	25	29
3.2	1.1632	1663	1694	1725	1756	1787	1817	1848	1878	1909	3	6	9	12	15	18	22	25	28
3.3	1.1939	1969	2000	2030	2060	2090	2119	2149	2179	2208	3	6	9	12	15	18	21	24	26
3.4	1.2238	2267	2296	2326	2355	2384	2413	2442	2470	2499	3	6	9	12	15	17	20	23	26
3.5	1.2528	2556	2585	2613	2641	2669	2698	2726	2754	2782	3	6	8	11	14	17	20	23	25
3.6	1.2809	2837	2865	2892	2920	2947	2975	3002	3029	3056	3	5	8	11	14	16	19	22	25
3.7	1.3083	3110	3137	3164	3191	3218	3244	3271	3297	3324	3	5	8	11	13	16	19	21	24
3.8	1.3350	3376	3403	3429	3455	3481	3507	3533	3558	3584	3	5	8	10	13	16	18	21	23
3.9	1.3610	3635	3661	3686	3712	3737	3762	3788	3813	3838	3	5	8	10	13	15	18	20	23
4.0	1.3863	3888	3913	3938	3962	3987	4012	4036	4061	4085	2	5	7	10	12	15	17	20	22
4.1	1.4110	4134	4159	4183	4207	4231	4255	4279	4303	4327	2	5	7	10	12	14	17	19	22
4.2	1.4351	4375	4398	4422	4446	4469	4493	4516	4540	4563	2	5	7	9	12	14	16	19	21
4.3	1.4586	4609	4633	4656	4679	4702	4725	4748	4770	4793	2	5	7	9	12	14	16	18	21
4.4	1.4816	4839	4861	4884	4907	4929	4951	4974	4996	5019	2	4	7	9	11	13	16	18	20
4.5	1.5041	5063	5085	5107	5129	5151	5173	5195	5217	5239	2	4	7	9	11	13	15	18	20
4.6	1.5261	5282	5304	5326	5347	5369	5390	5412	5433	5454	2	4	6	9	11	13	15	17	19
4.7	1.5476	5497	5518	5539	5560	5581	5602	5623	5644	5665	2	4	6	8	11	13	15	17	19
4.8	1.5686	5707	5728	5748	5769	5790	5810	5831	5851	5872	2	4	6	8	10	12	14	16	19
4.9	1.5892	5913	5933	5953	5974	5994	6014	6034	6054	6074	2	4	6	8	10	12	14	16	18

x	$\log_e 10^{-x}$
1	$\overline{3}$.6974
2	$\overline{5}$.3948
3	$\overline{7}$.0922
4	$\overline{10}$.7897
5	$\overline{12}$.4871
6	$\overline{14}$.1845
7	$\overline{17}$.8819
8	$\overline{19}$.5793
9	$\overline{21}$.2767
10	$\overline{24}$.3741

x	0	1	2	3	4	5	6	7	8	9	1	2	3	4	5	6	7	8	9
											ADD PROPORTIONAL PARTS								
5.0	1.6094	6114	6134	6154	6174	6194	6214	6233	6253	6273	2	4	6	8	10	12	14	16	18
5.1	1.6292	6312	6332	6351	6371	6390	6409	6429	6448	6467	2	4	6	8	10	12	14	16	17
5.2	1.6487	6506	6525	6544	6563	6582	6601	6620	6639	6658	2	4	6	8	10	11	13	15	17
5.3	1.6677	6696	6715	6734	6752	6771	6790	6808	6827	6845	2	4	6	8	9	11	13	15	17
5.4	1.6864	6882	6901	6919	6938	6956	6974	6993	7011	7029	2	4	6	7	9	11	13	15	17
5.5	1.7047	7066	7084	7102	7120	7138	7156	7174	7192	7210	2	4	5	7	9	11	13	14	16
5.6	1.7228	7246	7263	7281	7299	7317	7334	7352	7370	7387	2	4	5	7	9	11	12	14	16
5.7	1.7405	7422	7440	7457	7475	7492	7509	7527	7544	7561	2	3	5	7	9	10	12	14	16
5.8	1.7579	7596	7613	7630	7647	7664	7681	7699	7716	7733	2	3	5	7	9	10	12	14	15
5.9	1.7750	7766	7783	7800	7817	7834	7851	7867	7884	7901	2	3	5	7	8	10	12	13	15
6.0	1.7918	7934	7951	7967	7984	8001	8017	8034	8050	8066	2	3	5	7	8	10	12	13	15
6.1	1.8083	8099	8116	8132	8148	8165	8181	8197	8213	8229	2	3	5	6	8	10	11	13	15
6.2	1.8245	8262	8278	8294	8310	8326	8342	8358	8374	8390	2	3	5	6	8	9	11	13	14
6.3	1.8405	8421	8437	8453	8469	8485	8500	8516	8532	8547	2	3	5	6	8	9	11	13	14
6.4	1.8563	8579	8594	8610	8625	8641	8656	8672	8687	8703	2	3	5	6	8	9	11	12	14
6.5	1.8718	8733	8749	8764	8779	8795	8810	8825	8840	8856	2	3	5	6	8	9	11	12	14
6.6	1.8871	8886	8901	8916	8931	8946	8961	8976	8991	9006	1	3	4	6	8	9	11	12	14
6.7	1.9021	9036	9051	9066	9081	9095	9110	9125	9140	9155	1	3	4	6	7	9	10	12	13
6.8	1.9169	9184	9199	9213	9228	9242	9257	9272	9286	9301	1	3	4	6	7	9	10	12	13
6.9	1.9315	9330	9344	9359	9373	9387	9402	9416	9430	9445	1	3	4	6	7	9	10	11	13
7.0	1.9459	9473	9488	9502	9516	9530	9544	9559	9573	9587	1	3	4	6	7	9	10	11	13
7.1	1.9601	9615	9629	9643	9657	9671	9685	9699	9713	9727	1	3	4	6	7	8	10	11	13
7.2	1.9741	9755	9769	9782	9796	9810	9824	9838	9851	9865	1	3	4	5	7	8	10	11	12
7.3	1.9879	9892	9906	9920	9933	9947	9961	9974	9988	0001	1	3	4	5	7	8	10	11	12
7.4	2.0015	0028	0042	0055	0069	0082	0096	0109	0122	0136	1	3	4	5	7	8	9	11	12
7.5	2.0149	0162	0176	0189	0202	0215	0229	0242	0255	0268	1	3	4	5	7	8	9	11	12
7.6	2.0281	0295	0308	0321	0334	0347	0360	0373	0386	0399	1	3	4	5	6	8	9	10	12
7.7	2.0412	0425	0438	0451	0464	0477	0490	0503	0516	0528	1	3	4	5	6	8	9	10	12
7.8	2.0541	0554	0567	0580	0592	0605	0618	0631	0643	0656	1	3	4	5	6	8	9	10	11
7.9	2.0669	0681	0694	0707	0719	0732	0744	0757	0769	0782	1	3	4	5	6	8	9	10	11
8.0	2.0794	0807	0819	0832	0844	0857	0869	0882	0894	0906	1	2	4	5	6	7	9	10	11
8.1	2.0919	0931	0943	0956	0968	0980	0992	1005	1017	1029	1	2	4	5	6	7	8	10	11
8.2	2.1041	1054	1066	1078	1090	1102	1114	1126	1138	1150	1	2	4	5	6	7	8	10	11
8.3	2.1163	1175	1187	1199	1211	1223	1235	1247	1258	1270	1	2	4	5	6	7	8	10	11
8.4	2.1282	1294	1306	1318	1330	1342	1353	1365	1377	1389	1	2	4	5	6	7	8	9	11
8.5	2.1401	1412	1424	1436	1448	1459	1471	1483	1494	1506	1	2	4	5	6	7	8	9	11
8.6	2.1518	1529	1541	1552	1564	1576	1587	1599	1610	1622	1	2	3	5	6	7	8	9	10
8.7	2.1633	1645	1656	1668	1679	1691	1702	1713	1725	1736	1	2	3	5	6	7	8	9	10
8.8	2.1748	1759	1770	1782	1793	1804	1815	1827	1838	1849	1	2	3	4	6	7	8	9	10
8.9	2.1861	1872	1883	1894	1905	1917	1928	1939	1950	1961	1	2	3	4	6	7	8	9	10
9.0	2.1972	1983	1994	2006	2017	2028	2039	2050	2061	2072	1	2	3	4	5	7	8	9	10
9.1	2.2083	2094	2105	2116	2127	2138	2148	2159	2170	2181	1	2	3	4	5	6	8	9	10
9.2	2.2192	2203	2214	2225	2235	2246	2257	2268	2279	2289	1	2	3	4	5	6	8	9	10
9.3	2.2300	2311	2322	2332	2343	2354	2364	2375	2386	2396	1	2	3	4	5	6	7	9	10
9.4	2.2407	2418	2428	2439	2450	2460	2471	2481	2492	2502	1	2	3	4	5	6	7	8	10
9.5	2.2513	2523	2534	2544	2555	2565	2576	2586	2597	2607	1	2	3	4	5	6	7	8	9
9.6	2.2618	2628	2638	2649	2659	2670	2680	2690	2701	2711	1	2	3	4	5	6	7	8	9
9.7	2.2721	2732	2742	2752	2762	2773	2783	2793	2803	2814	1	2	3	4	5	6	7	8	9
9.8	2.2824	2834	2844	2854	2865	2875	2885	2895	2905	2915	1	2	3	4	5	6	7	8	9
9.9	2.2925	2935	2946	2956	2966	2976	2986	2996	3006	3016	1	2	3	4	5	6	7	8	9

x	$\log_e 10^{x}$
1	2.3026
2	4.6052
3	6.9078
4	9.2103
5	11.5129
6	13.8155
7	16.1181
8	18.4207
9	20.7233
10	23.0259

Common logarithms: $\log_{10} x$

x	0	1	2	3	4	5	6	7	8	9	1	2	3	4	5	6	7	8	9
10	0000	0043	0086	0128	0170	0212	0253	0294	0334	0374	4	9	13	17	21	25	30	34	38
11	0414	0453	0492	0531	0569	0607	0645	0682	0719	0755	4	8	12	16	20	24	28	32	36
12	0792	0828	0864	0899	0934	0969	1004	1038	1072	1106	4	8	11	15	19	23	27	31	35
13	1139	1173	1206	1239	1271	1303	1335	1367	1399	1430	3	7	10	14	17	21	24	28	31
14	1461	1492	1523	1553	1584	1614	1644	1673	1703	1732	3	6	10	13	16	19	23	26	30
15	1761	1790	1818	1847	1875	1903	1931	1959	1987	2014	3	6	9	11	14	17	20	23	26
16	2041	2068	2095	2122	2148	2175	2201	2227	2253	2279	3	5	8	11	14	16	19	22	25
17	2304	2330	2355	2380	2405	2430	2455	2480	2504	2529	3	5	8	10	13	15	18	20	23
18	2553	2577	2601	2625	2648	2672	2695	2718	2742	2765	2	5	7	10	12	14	17	19	21
19	2788	2810	2833	2856	2878	2900	2923	2945	2967	2989	2	5	7	9	12	14	16	18	20
20	3010	3032	3054	3075	3096	3118	3139	3160	3181	3201	2	4	6	8	11	13	15	17	19
21	3222	3243	3263	3284	3304	3324	3345	3365	3385	3404	2	4	6	8	10	12	14	16	18
22	3424	3444	3464	3483	3502	3522	3541	3560	3579	3598	2	4	6	8	10	12	14	15	17
23	3617	3636	3655	3674	3692	3711	3729	3747	3766	3784	2	4	6	7	9	11	13	15	17
24	3802	3820	3838	3856	3874	3892	3909	3927	3945	3962	2	4	5	7	9	11	12	14	16
25	3979	3997	4014	4031	4048	4065	4082	4099	4116	4133	2	3	5	7	9	10	12	14	15
26	4150	4166	4183	4200	4216	4232	4249	4265	4281	4298	2	3	5	7	8	10	11	13	15
27	4314	4330	4346	4362	4378	4393	4409	4425	4440	4456	2	3	5	6	8	9	11	13	14
28	4472	4487	4502	4518	4533	4548	4564	4579	4594	4609	2	3	5	6	8	9	11	12	14
29	4624	4639	4654	4669	4683	4698	4713	4728	4742	4757	1	3	4	6	7	9	10	12	13
30	4771	4786	4800	4814	4829	4843	4857	4871	4886	4900	1	3	4	6	7	9	10	11	13
31	4914	4928	4942	4955	4969	4983	4997	5011	5024	5038	1	3	4	6	7	8	10	11	12
32	5051	5065	5079	5092	5105	5119	5132	5145	5159	5172	1	3	4	5	7	8	9	11	12
33	5185	5198	5211	5224	5237	5250	5263	5276	5289	5302	1	3	4	5	6	8	9	10	12
34	5315	5328	5340	5353	5366	5378	5391	5403	5416	5428	1	2	4	5	6	7	9	10	11
35	5441	5453	5465	5478	5490	5502	5514	5527	5539	5551	1	2	4	5	6	7	8	10	11
36	5563	5575	5587	5599	5611	5623	5635	5647	5658	5670	1	2	4	5	6	7	8	9	11
37	5682	5694	5705	5717	5729	5740	5752	5763	5775	5786	1	2	3	5	6	7	8	9	10
38	5798	5809	5821	5832	5843	5855	5866	5877	5888	5899	1	2	3	5	6	7	8	9	10
39	5911	5922	5933	5944	5955	5966	5977	5988	5999	6010	1	2	3	4	5	7	8	9	10
40	6021	6031	6042	6053	6064	6075	6085	6096	6107	6117	1	2	3	4	5	6	7	9	10
41	6128	6138	6149	6160	6170	6180	6191	6201	6212	6222	1	2	3	4	5	6	7	8	9
42	6232	6243	6253	6263	6274	6284	6294	6304	6314	6325	1	2	3	4	5	6	7	8	9
43	6335	6345	6355	6365	6375	6385	6395	6405	6415	6425	1	2	3	4	5	6	7	8	9
44	6435	6444	6454	6464	6474	6484	6493	6503	6513	6522	1	2	3	4	5	6	7	8	9
45	6532	6542	6551	6561	6571	6580	6590	6599	6609	6618	1	2	3	4	5	6	7	8	9
46	6628	6637	6646	6656	6665	6675	6684	6693	6702	6712	1	2	3	4	5	6	6	7	8
47	6721	6730	6739	6749	6758	6767	6776	6785	6794	6803	1	2	3	4	4	5	6	7	8
48	6812	6821	6830	6839	6848	6857	6866	6875	6884	6893	1	2	3	4	4	5	6	7	8
49	6902	6911	6920	6928	6937	6946	6955	6964	6972	6981	1	2	3	4	4	5	6	7	8
x	0	1	2	3	4	5	6	7	8	9	1	2	3	4	5	6	7	8	9

ADD PROPORTIONAL PARTS

x	0	1	2	3	4	5	6	7	8	9	1	2	3	4	5	6	7	8	9
50	6990	6998	7007	7016	7024	7033	7042	7050	7059	7067	1	2	3	3	4	5	6	7	8
51	7076	7084	7093	7101	7110	7118	7126	7135	7143	7152	1	2	3	3	4	5	6	7	8
52	7160	7168	7177	7185	7193	7202	7210	7218	7226	7235	1	2	2	3	4	5	6	6	7
53	7243	7251	7259	7267	7275	7284	7292	7300	7308	7316	1	2	2	3	4	5	6	6	7
54	7324	7332	7340	7348	7356	7364	7372	7380	7388	7396	1	2	2	3	4	5	6	6	7
55	7404	7412	7419	7427	7435	7443	7451	7459	7466	7474	1	2	2	3	4	5	5	6	7
56	7482	7490	7497	7505	7513	7520	7528	7536	7543	7551	1	2	2	3	4	5	5	6	7
57	7559	7566	7574	7582	7589	7597	7604	7612	7619	7627	1	2	2	3	4	5	5	6	7
58	7634	7642	7649	7657	7664	7672	7679	7686	7694	7701	1	1	2	3	4	4	5	6	7
59	7709	7716	7723	7731	7738	7745	7752	7760	7767	7774	1	1	2	3	4	4	5	6	7
60	7782	7789	7796	7803	7810	7818	7825	7832	7839	7846	1	1	2	3	4	4	5	6	6
61	7853	7860	7868	7875	7882	7889	7896	7903	7910	7917	1	1	2	3	4	4	5	6	6
62	7924	7931	7938	7945	7952	7959	7966	7973	7980	7987	1	1	2	3	4	4	5	6	6
63	7993	8000	8007	8014	8021	8028	8035	8041	8048	8055	1	1	2	3	3	4	5	6	6
64	8062	8069	8075	8082	8089	8096	8102	8109	8116	8122	1	1	2	3	3	4	5	5	6
65	8129	8136	8142	8149	8156	8162	8169	8176	8182	8189	1	1	2	3	3	4	5	5	6
66	8195	8202	8209	8215	8222	8228	8235	8241	8248	8254	1	1	2	3	3	4	5	5	6
67	8261	8267	8274	8280	8287	8293	8299	8306	8312	8319	1	1	2	3	3	4	5	5	6
68	8325	8331	8338	8344	8351	8357	8363	8370	8376	8382	1	1	2	3	3	4	4	5	6
69	8388	8395	8401	8407	8414	8420	8426	8432	8439	8445	1	1	2	2	3	4	4	5	6
70	8451	8457	8463	8470	8476	8482	8488	8494	8500	8506	1	1	2	2	3	4	4	5	5
71	8513	8519	8525	8531	8537	8543	8549	8555	8561	8567	1	1	2	2	3	4	4	5	5
72	8573	8579	8585	8591	8597	8603	8609	8615	8621	8627	1	1	2	2	3	4	4	5	5
73	8633	8639	8645	8651	8657	8663	8669	8675	8681	8686	1	1	2	2	3	4	4	5	5
74	8692	8698	8704	8710	8716	8722	8727	8733	8739	8745	1	1	2	2	3	4	4	5	5
75	8751	8756	8762	8768	8774	8779	8785	8791	8797	8802	1	1	2	2	3	3	4	5	5
76	8808	8814	8820	8825	8831	8837	8842	8848	8854	8859	1	1	2	2	3	3	4	5	5
77	8865	8871	8876	8882	8887	8893	8899	8904	8910	8915	1	1	2	2	3	3	4	4	5
78	8921	8927	8932	8938	8943	8949	8954	8960	8965	8971	1	1	2	2	3	3	4	4	5
79	8976	8982	8987	8993	8998	9004	9009	9015	9020	9025	1	1	2	2	3	3	4	4	5
80	9031	9036	9042	9047	9053	9058	9063	9069	9074	9079	1	1	2	2	3	3	4	4	5
81	9085	9090	9096	9101	9106	9112	9117	9122	9128	9133	1	1	2	2	3	3	4	4	5
82	9138	9143	9149	9154	9159	9165	9170	9175	9180	9186	1	1	2	2	3	3	4	4	5
83	9191	9196	9201	9206	9212	9217	9222	9227	9232	9238	1	1	2	2	3	3	4	4	5
84	9243	9248	9253	9258	9263	9269	9274	9279	9284	9289	1	1	2	2	3	3	4	4	5
85	9294	9299	9304	9309	9315	9320	9325	9330	9335	9340	1	1	2	2	3	3	4	4	5
86	9345	9350	9355	9360	9365	9370	9375	9380	9385	9390	1	1	2	2	3	3	4	4	5
87	9395	9400	9405	9410	9415	9420	9425	9430	9435	9440	0	1	1	2	2	3	3	4	4
88	9445	9450	9455	9460	9465	9469	9474	9479	9484	9489	0	1	1	2	2	3	3	4	4
89	9494	9499	9504	9509	9513	9518	9523	9528	9533	9538	0	1	1	2	2	3	3	4	4
90	9542	9547	9552	9557	9562	9566	9571	9576	9581	9586	0	1	1	2	2	3	3	4	4
91	9590	9595	9600	9605	9609	9614	9619	9624	9628	9633	0	1	1	2	2	3	3	4	4
92	9638	9643	9647	9652	9657	9661	9666	9671	9675	9680	0	1	1	2	2	3	3	4	4
93	9685	9689	9694	9699	9703	9708	9713	9717	9722	9727	0	1	1	2	2	3	3	4	4
94	9731	9736	9741	9745	9750	9754	9759	9763	9768	9773	0	1	1	2	2	3	3	4	4
95	9777	9782	9786	9791	9795	9800	9805	9809	9814	9818	0	1	1	2	2	3	3	4	4
96	9823	9827	9832	9836	9841	9845	9850	9854	9859	9863	0	1	1	2	2	3	3	4	4
97	9868	9872	9877	9881	9886	9890	9894	9899	9903	9908	0	1	1	2	2	3	3	4	4
98	9912	9917	9921	9926	9930	9934	9939	9943	9948	9952	0	1	1	2	2	3	3	4	4
99	9956	9961	9965	9969	9974	9978	9983	9987	9991	9996	0	1	1	2	2	3	3	4	4
x	0	1	2	3	4	5	6	7	8	9	1	2	3	4	5	6	7	8	9

ADD PROPORTIONAL PARTS